Aspects of Personal Privacy in Communications: Problems, Technology and Solutions

RIVER PUBLISHERS SERIES IN COMMUNICATIONS

This series focuses on communications science and technology. This includes the theory and use of systems involving all terminals, computers, and information processors; wired and wireless networks; and network layouts, procontentsols, architectures, and implementations.

Furthermore, developments toward new market demands in systems, products, and technologies such as personal communications services, multimedia systems, enterprise networks, and optical communications systems.

- Wireless Communications
- Networks
- Security
- Antennas & Propagation
- Microwaves
- Software Defined Radio

For a list of other books in this series, visit
http://riverpublishers.com/river_publisher/series.php?msg=Communications

Aspects of Personal Privacy in Communications: Problems, Technology and Solutions

Geir M. Køien
University of Agder, Norway

and

Vladimir A. Oleshchuk
University of Agder, Norway

River Publishers

Routledge
Taylor & Francis Group
LONDON AND NEW YORK

Published 2013 by River Publishers
River Publishers
Alsbjergvej 10, 9260 Gistrup, Denmark
www.riverpublishers.com

Distributed exclusively by Routledge
4 Park Square, Milton Park, Abingdon, Oxon OX14 4RN
605 Third Avenue, New York, NY 10158

First published in paperback 2024

Aspects of Personal Privacy in Communications: Problems, Technology and Solutions / by Geir M. Køien, Vladimir A. Oleshchuk.

Routledge is an imprint of the Taylor & Francis Group, an informa business

ISBN: 978-87-92982-08-7 (hbk)
ISBN: 978-87-7004-509-4 (pbk)

Contents

Preface

The modern society is rapidly becoming a fully digital society. Our everyday activities are increasingly being monitored and recorded. The registrations include data concerning our buying habits, medical records, reading, and viewing preferences, and traveling habits. The collected data are recorded and analyzed in detail. The growing number of personal devices and gadgets that we use emphasizes the scale of the digitalizing of our lives.

Our personal privacy is threatened. The threat does not so much come from a 1984 style Big Brother, but rather from a set of smaller big brothers. The small big brothers are companies that we interact with; they are public services and institutions. Many of these small big brothers are indeed also being invited to our private data by ourselves. By giving up a little privacy we may also stand to gain, but do we know the trade-offs? There are benefits in the form of convenience and improvement of personal safety, security and quality of provided services. The benefits can also transcend the individual and be of benefit to society if one is able to use the accumulated private information wisely.

However, such services may invade people's privacy and there clearly is a trade-off between safety/security and privacy in the sense that we may have to give up some personal privacy in exchange for better safety/security. Here, like in so many other circumstances, there is an inherent asymmetry in who stands to gain and who must give the most. Companies and large public institutions don't take the individuals stand unless there either is something to be gained (reputation and customer satisfaction etc) or they are forced to respect our privacy due to regulations and laws.

Privacy as a subject can be problematic. At the extreme, it is personal freedom against safety and security. We shall not take a political stand on personal privacy and what level of personal freedom and privacy is the correct one.

This book is mostly about understanding what privacy is and some of the technologies may help us to regain a bit of privacy. We discuss what privacy is about, what the different aspects of privacy may be and why privacy needs to be there by default. There are boundaries between personal privacy and societal requirements, and inevitably society will set limits to our privacy (Lawful Interception etc.). There are technologies that are specifically designed to help us regain some digital privacy. These are commonly known as Privacy Enhancing Technologies (PETs). We investigate some of these PETs including MIX networks, Onion Routing and various privacy-preserving methods. Other aspects include identity- and location privacy in cellular systems, privacy in RFID, Internet-of-Things (IoT) and sensor networks amongst others. Some aspects of cloud systems are also covered.

Reading the book

Part I of this book is relatively high level. An interest in technological matters should suffice. Part II of the book is quite technical and some sections require a certain level of mathematical sophistication in order to be fully understood. The tougher going sections may safely be skipped without causing problems for understanding later sections. Part III should be accessible to most readers with a technical background.

Caveat Emptor

While we do discuss legal aspects of privacy in this book, we must confess that we are not lawyers. We also discuss various societal aspects of privacy in the book, but we are just concerned individuals and have no formal education in these aspects. We hope this helps.

Acknowledgment

Patience is a virtue. We hereby thank those around us and the staff at River Publishers for being patient with us.

Geir M. Køien and **Vladimir A. Oleshcuk**

Grimstad, Norway, 2013

Part I

The Scope of Privacy

1

Getting a Grip on Privacy

Welcome to the first chapter of the book. In this chapter we aim to define the scope of the book and introduce a number of concepts and notions. Much of the content of this book is highly technical, but the moral and ethical justifications for privacy and the societal and personal arguments for and against privacy are not technical in nature. We start off with an outline of the philosophical *Right to Privacy*.

From that we discuss exactly what privacy is. Our notion of privacy is in many respects a function of our culture and the values of our society. Making it worse, it is a fluid concept and a concept that changes over time.

In section 1.2 "Privacy Matters" we continue our investigation about what privacy really is and how digital technology affects our privacy. We also define some central and crucial aspects of privacy. In the next subsections, we investigate the concept of *Identifiers*. And, what is an identifier and what is an identity? People, of course, have names and names are good references (read: identifiers) to other people, but names are not the only identity or reference to you and in a digital world it is not the most important one either.

Identity concealment is a central part of many privacy preserving techniques, but identity is not the only privacy attribute that needs attention. In section 1.4 "Aspects of Privacy", we look into some types of privacy. We then proceed with practical approaches towards privacy. With respect to the practical part, we have the so-called Privacy Enhancing Technologies (PETs) approach. The PETs are specific technological methods for archiving specific goals. While PETs may provide practical tools, the general approach has been criticized and it has even been suggested that PETs may be a hindrance to achieving fuller and more complete privacy [235]. Moving on, we investigate the *Privacy by Design* (PbD) approach. The PbD manifesto is partly philosophical, partially policy-based, partly activist and partly practical in terms of providing privacy as the ground condition for all ICT systems. The context for the policy part of the PbD paradigm is wide, and this may indeed provide an answer to the problem of limited scope that the PETs-only approach faces.

When it comes to privacy technology terminology definitions, we recommend the so-called "Anonymity Terminology" report [211], but we will still often prefer to use intuitive everyday terms.

1.1 The Right to Privacy

Privacy and our assumed right to privacy is nothing new. Privacy as a concept is, of course, independent of digital technology. The concept of privacy as such seems to be on par with concepts such as *freedom*, *security* (in a physical sense) and (personal) *safety*.

In the essay "The Right to Privacy" [248], the authors give a broad and compelling range of arguments for mans right to personal privacy. The authors highlight the increasing capacity of government, the press, and other agencies and institutions to invade our privacy. The authors argue that protection of the private realm is the essence of individual freedom in the modern age. The authors published the above essay in late 1890 and so the "*modern age*" they relate to isn't the digital world of today. They also did not write the article with a technical angle; In fact, the essay was published in Harvard Law Review. Not surprisingly then, the authors argue that the law must evolve in response to technological change.

Among the diverse set of privacy rights discussed is the right "*to be let alone*". The authors continue to declare that "*...the individual is entitled to decide whether that which is his shall be given to the public.*". The arguments go beyond aspects such as copyrights or property rights and focus on privacy as a generic right not to disclose personal information without due consent.

The authors write from a perspective of generic legal rights and given that the article was published in 1890 clearly does not take into account technical aspects of privacy in a digital context. However, the authors foresaw the need to capture privacy as a generic right and declared that the right applies to "*any modern device*" that can infringe on the personal privacy rights.

To round off, the authors recognized that "4. The right to privacy ceases upon the publication of the facts by the individual, or with his consent.". We note that this particular aspect may need qualification in a digital world. Indeed, there seems to be a consensus amongst privacy agencies, privacy ombudsmen and indeed in some part of the industry that there also exists a right "*to be forgotten*". That is, if you as a private person publish some silly pictures of your self, you have a certain right to annul the pictures. Certainly, you can delete and remove the picture from your website, but the right also extends to a right to be forgotten by search engines. There is no universal consensus, but it seems that many jurisdictions in Europe consider 3– 5 years a suitable period, after which privately published data should no longer be accessible by search engines. The right "*to be forgotten*" is not without controversies, and, in particular, it may go against the right to "*free speech*". This aspect seems particularly expressed for content that was not published in a private context (magazines, newspapers), but which is about private persons. What right do we, as private citizens, have to be forgotten about if the offensive material was published by a newspaper? There is also a distinction to be made and a line to be drawn when it comes to what we mean by "private citizen". Clearly, the president of the U.S.A. cannot expect the same right to be forgotten about as does Average Joe, but where should the line be drawn?

1.2 Privacy Matters

Accepting that we have a right "to be let alone", how do we capture what this right is all about? What properties does privacy have? In the following, we further investigate privacy as a concept and try to define it in different contexts.

1.2.1 An Intrinsic Property of Privacy

Privacy has the property that once a privacy sensitive piece of information is exposed then the privacy is lost, potentially forever. Thus, we have the "*once exposed - forever lost*" maxim of privacy.

Of course, the maxim does not necessarily have to be true, but in a worst-case sense one must assume it to be true. People forget and forgive, information is lost, what once was important becomes irrelevant and the past fades into oblivion. At least this was the way things used to be.

Today, in a digital and more-or-less fully connected world, much can be known about us and little need to be lost. Digital information is leaked and exposed in many subtle ways, and with this in mind it is tempting to view *loss of privacy* as congruent with the concept of entropy[1] (i.e. as an entity that always and invariably increases), but that may turn out to be an invalid analogy.

[1] We are thinking mainly of an information theoretical notion of entropy here, but a thermodynamical definition will also suit the purpose.

Theoretically, in a worst-case sense, the exposed/leaked information may forever stay exposed, so the analogy seem fair. There are also, in an information theoretical sense, strong indications that information will indeed leak over time. So, why may the analogy be false? Well, privacy loss is a notion that that is related not only to the fact that some information is exposed, but also to the privacy sensitive property of the information. And this aspect of privacy is not so much subject to information theory as to social norms, religious beliefs, moral standards, cultural values etc. And, in a real-time sense, all information, privacy sensitive or not, may be viewed as having a time-dependent validity period. Put another way, the validity, or importance, of a piece of information may *age* and become less relevant over time. Given a change of context the opposite may, of course, also be true; and people that suddenly become famous will discover that what once was irrelevant details from their past suddenly has got news value, and an embarrassment that was all but forgotten is suddenly revived and more painfully embarrassing than ever. Anyway, the amount of lost privacy does not have to monotonically increase, but may actually decrease over time if the information becomes less privacy sensitive.

Another matter is that much information is actually lost or has become inaccessible. It is hard to estimate the rate of the loss, but digital information is not immune to deletions and even when not deleted, it may over time be impossible to interpret the information correctly. The loss and deterioration of stored data/information is due to to range of different reasons, including physical decay of the storage media, inability to read off the physical media, inability to decode that raw data, and inability to correctly interpret the data once decoded.

An simple example might be a WordStar text document created for the CP/M operating system and stored on a $5\frac{1}{4}$ inch diskette. Consider then that the file contained information about a minor crime committed by a farmer living on the north-east coast of Iceland in 1987. Obviously, the text document would be written in Icelandic. How many would, without spending a lot of effort, be able to read the data and understand the information today? The same issues and more are true for digital libraries and the task of digital preservation. The articles in [27, 82, 185] detail some of these issues, including authenticity, and demonstrate that it takes a lot of effort to avoid digital information rot.

The *once exposed - forever lost* maxim is still valid, but not without qualifications. However, for privacy we cannot rely on luck and forgetfulness, and so we must assume the maxim to be true.

1.2.2 Privacy Sensitive Information

Before we proceed, we need to have at least an informal definition about what we mean by "privacy sensitive". The following questions serve to highlight some aspects of privacy sensitive:

- What exactly is privacy sensitive information?
- When does some piece of information become privacy sensitive?
- Can information stop being privacy sensitive?

First, we observe that "privacy sensitive" is a relative concept. Specifically, a piece of information is privacy sensitive with respect to somebody (the subject/subjects). Secondly, what is considered privacy sensitive is also relative to the observers. Thirdly, we note that the sensitiveness is determined by the context and is influenced by individual preferences, social norms, religious beliefs, moral standards, cultural values, etc etc. Some examples are given below to clarify the concept:

- **Nudity**
  Our "private parts" is an obvious example. If you are visiting a nude beach, nudity is expected. In some cultures nudity is nothing special, but in most of the world nudity is regarded as private and privacy sensitive.
- **Religious beliefs**
  Religious beliefs may be privacy sensitive by themselves. There are also many other privacy sensitive aspects. If you are a believer then there are commandments and moral codes to adhere to. Knowledge about a believer not adhering to these codes may be considered privacy sensitive.
- **Promiscuity**
  Extramarital affairs and promiscuous behavior is often, but not always, very privacy sensitive. The sensitivity is linked to the amount of embarrassment and to the damage done if the secret is revealed.
- **Income**
  Our financial status and our income is often, but not always, privacy sensitive information.

We round off here by noting that privacy sensitive is a dynamic property. Some information, that previously was of little matter, may become quite sensitive. Remember the Bill Clinton quote "When I was in England, I experimented with marijuana a time or two, and I didn't like it. I didn't inhale and never tried it again.". It clearly was a major embarrassment for Clinton to admit smoking marijuana, but it would have been utterly unimportant if he had not been the president of the U.S.A. Of cours,e the opposite is also true.

People forgive and forget, and today it matters little to Clinton's reputation whether he inhaled or not.

1.2.3 Security, Safety and Privacy

In the following, we give brief, incomplete and somewhat inconsistent definitions of the terms *Security*, *Safety* and *Privacy*. We do this in order to convey a general notion of what we mean by these concepts in the context of this book. In normal everyday speech the concepts may seem blurry and overlapping, and to some extent they do overlap; or at least, they do interact with each other. On the other hand, there are important differences too, and one should be aware of these. With this clarifying disclaimer in mind we proceed with the definitions[2].

1.2.3.1 What do we mean by "Security"

Security can be defined as "the degree of protection against danger, damage, injury, or losses". One often defines this in a context of an adversary or intruder, but strictly speaking this definition does not require anyone with malicious intent.

One may also define security as "a form of protection where a separation is created between the assets and the threat". Here, one has defined assets that can be threatened, and security is then something which protects the assets and/or mitigates the threat. The definition still does not require an intruder or indeed any malicious intent. Many threats are, of course, posed by some kind of adversary. Malicious intent is still not needed, and frankly the concept is irrelevant in our context. It is better to use a somewhat more neutral term like "deliberate intent".

Security can also be defined and compared to the related concepts of *safety* and *reliability*. The main difference between security and reliability is that security must take into account a possible deliberate intention to threaten the given asset. An example of the difference is illustrated in how one defines integrity protection in communications and in security:

- **Communications**
  Integrity protection of a data packet is often by means of coding redundancy and by adding a cyclic redundancy check code. For instance, in the GSM system, one uses a scheme with combined block coding,

[2] The reader is invited to look up the terms in thesauruses and ponder the different definitions; while keeping the context in mind.

convolutional coding and block interleaving. This scheme is designed specifically to mitigate and detect the sort of errors that most frequently occur in GSM radio related transmissions. These errors may be complex, but there is no deliberate intent to them.

- **Security**
  Integrity protection of a data packet in a security context must take into account an intruder that deliberately wants to corrupt the data. The integrity protection offered by communications methods, sophisticated as they may be, offer no protection against wilful manipulations. Therefore, one typically adds a cryptographic checksum computed over all the content. The data may still be manipulated, but any modification would be detectable by the receiver[3].

Security may be considered a prerequisite for privacy, but security alone does not necessarily provide privacy. Indeed, sometimes security is at odds with privacy.

1.2.3.2 What do we mean by "Safety"

One sometimes needs a clear distinction between protection against accidents and protection against an adversary.

Safety is then used to define protection against failures, damages, errors, accidents etc. against physical, social, financial, emotional, psychological threats. Safety can also be captured as control of recognized hazards and threats to achieve an acceptable level of risk. It is still possible that a person caused the damage etc in question, but we then strongly assume that the person did not intend to cause any damage or disruption. Thus, we classify the incident as an accident and not an attack. To illustrate the differencc, wc have that regulations to fasten seat belts in cars are a safety measure, while metal detector screening at airports is a security measure.

1.2.3.3 What do we mean by "Privacy"

Privacy is the ability, of an individual or a small group of people, to restrict or prevent information about oneself to be revealed without consent.

So privacy is about control over information that is private. There is no universal accord of what is private in this respect; it differs over time, among cultures and individuals. There is nevertheless a common core of what is considered private. Privacy includes the ability to remain anonymous, unno-

[3] Only with access to the security credential (secret key etc) can one modify/verify the cryptographic checksum.

ticed or unidentified in a public context. Privacy is relative in this respect, as anonymity is with respect to someone or something. The main point is that one is able to control knowledge about, for instance, identity and presence.

When something is considered to be private to a person, it means the person regards the information to be of special value to him/her and that it is something which is perceived to be personally sensitive. The "privacy sensitive" attribute depends on how the information will be received if it is made public. Privacy can thus be defined in relation to the embarrassment caused, the damage to reputation or generally to the amount of emotional distress caused by revealing the privacy sensitive information.

Security services may be used to protect privacy sensitive information, but security services does not automatically provide privacy. An example is that most authentication protocols rely on identity presentation in the first phase of the protocol, and this may obviously be problematic if the identity is considered private information or if exposing the identity may facilitate tracking or if it may lead to exposure of other sensitive information.

1.2.4 Personal Privacy and Entities

The term *personal privacy* may seem like a bit of an oxymoron; superfluous at best, confusing at worst. However, the term personal privacy is sometimes used when one wants to emphasize the fact that this is about humans. Privacy, as a generic term, may refer to "entities" as opposed to humans.

The term *entity* is in common use in information and communication technology (ICT) texts, and it generally refers to some program/process executing on a computer/device/gadget. The entity has one or more referential identities and performs various actions. However, as an abstract term, the concept of an entity would also encompass us humans. Thus, from a system point of view, we are entities.

1.2.5 Life-Time Considerations

What is the life-time of privacy sensitivity? Below, we briefly investigate this to put some perspective of the time horizon. Privacy, in common with security, tends to operate with worst-case scenarios and this greatly influences the viewpoints.

1.2.5.1 Data Storage

Privacy sensitive digital data needs to be stored in a protected manner. The lifetime of the data can be longer than the lifetime of the storage media. The protected data will need to be migrated onto new storage media during its lifetime. The life-time of the data may also outlast the cryptographic protection methods and the related key material. Thus, it must be possible to refresh the key material, negotiate new algorithms and cryptographic primitives for the protected data.

1.2.5.2 Archival Storage

Archival storage is where the lifespan is assumed to be very long (≥ 30 years). We assume that archival data objects will need to migrate between storage media and that protection methods will change during its lifespan.

1.2.5.3 Personal Privacy

The lifetime of a privacy object could potentially be that of the data subject itself. Potentially, the life-time is as long as the object which the sensitivity relates to; in short, it may ultimately be the life-time of the person in question [202]. The sensitivity may change over time, and much information will surely become less sensitive over time. However, we cannot rely on this and protection methods and schemes should be designed such that they potentially can protect the data for the life-time of a person.

1.3 Identifiers

Digital privacy is often about identifying or recognizing persons or entities. With respect to privacy, the questions is not so much "who" I am as "how" I am identified. Thus, we can avoid the philosophically challenging question of "who" and instead focus on "how" we identify a person. Sometimes the question is not "who", but rather "what". Thus, there are identifiers that primarily describe the ability or function of an individual, rather than identifying a specific individual.

1.3.1 Identities

There are many different definitions of "identity". We shall generally assume that an identity is completely capturing the defining characteristics of an object or entity. Thus, one may in some sense say that the identity *is* the object/entity. This means that an identity encompasses the coordinates (space/time) of an object in addition to actually being the object. In this sense, an exact replica is not considered the same as original. We do not suggest that the space/time coordinates should stay fixed for an entity; but that when the entity moves then it should still essentially remain the same.

1.3.2 Names and Aliases

One may define a name to be a word or phrase that is a distinctive designation of a person or thing. In a digital world, it may denote any type of object, and as such provides a reference to an entity.

We also have that an entity may have multiple names. These names need not be universally recognized, and they need not even be recognized by the named identity. The term "alias", which is derived from Latin (meaning "other"), is often used to denote secondary names. This, of course, presumes that one has a primary or main name.

A name may be a self-contained unique reference to an entity, but quite often it is only unique in a given context. Many people may have the name James Bond, but in the context of a fictional English MI6 agent the name will also be a unique reference. The context is then said to "qualify" the name, and one sometimes uses the term "fully qualified name/address" to indicate that the name/address is unique.

A name may also be considered a group name, and as such it references a set of entities. Your surname may, if one doesn't push the analogy too far, be considered a name for your family group. Names may also be private in the sense that one doesn't always want indiscriminate disclosure, and obviously a name may be publicly known. Names may also be transient and may cease to be while the object/entity still exists.

1.3.3 Links, Addresses and References

In our context, a link or a pointer is something which points to an address. The link or pointer is in many contexts directly denoting an address. For instance, in a program one may have a pointer `bref` with the value `0x337CFC00AB12`. `bref` is then used to indicate where (the address) the object/entity is stored in the computer memory. A reference may be referring to any kind of object, including itself or other references. An address is a physical/logical location. All objects/entities must somehow occupy a location, and the address then refers to where the entity is. A name can be considered a textual reference to an entity. Links or pointers are numerical indications to an entity. Of course, a link/pointer may be named. Note that the distinction between a name and a link is not always clear, and one may have numerical names that are also a link or pointer to an address and vice versa. The advantage to having names that aren't links/pointers is that they afford a level of indirection and thereby permit decoupling.

1.3.4 The Context as an Identifier

The context may itself serve as an identifier. That is, the information available during a transaction may in itself contain sufficient information to create an ad-hoc emergent identity. This may perhaps sound like a theoretical issue, but the Electronic Frontier Foundation (EFF) has made a neat illustration of this on their PanoptiClick homepage (`http://panopticlick.eff.org/`), where they demonstrate just how much context a web browser will present to a visited server. EFF has also made available an accompanying article on browser uniqueness [87]. Identification by context is, therefore, not a far fetched claim, but rather a reality under many circumstances.

1.3.5 Type Identifiers

A sufficiently complete description of an entity may be considered equivalent to identifying the entity. Thus, one may derive new alias identifiers.

However, sometimes it is more important to identify what you are than who you are. That is, to know the traits and the functions of an entity is sometimes more important than knowing its name. For instance, in a military battle it is important to know whether a solider is friend or foe; knowing his/her name becomes less important. Or, to know the licence plate number of a vehicle is a different matter from knowing whether the vehicle is an off-road SUV, an ordinary family car or a lorry. Type identifiers may typically be included as part of compound identifier.

1.3.6 Trust, Privacy and Identity

In the paper *Laws of Identity* [53], Kim Cameron set out to define the starting point for an *identity layer* for the internet. His observation was that the internet was built without a way to know who and what you are connecting to. So whenever we connect to the internet we potentially expose ourselves to dangers. And clearly there is scam, theft and deception carried out online these days and it does in many ways undermine public trust and usefulness of the internet. The *Laws of Identity* are not so much laws as they are directions for defining "a unifying identity metasystem that can offer the Internet the identity layer it so obviously requires". Without further ado, we here briefly present the *Laws of Identity*:

1. **User Control and Consent**
   Technical identity systems must only reveal information identifying a user with the users consent.
2. **Minimal Disclosure for a Constrained Use**
   The solution which discloses the least amount of identifying information and best limits its use is the most stable long term solution.
3. **Justifiable Parties**
   Digital identity systems must be designed so the disclosure of identifying information is limited to parties having a necessary and justifiable place in a given identity relationship.
4. **Directed Identity**
   A universal identity system must support both "omni-directional" identifiers for use by public entities and "unidirectional" identifiers for use by private entities, thus facilitating discovery while preventing unnecessary release of correlation handles.
5. **Pluralism of Operators and Technologies**
   A universal identity system must channel and enable the inter-working of multiple identity technologies run by multiple identity providers.
6. **Human Integration**
   The universal identity metasystem must define the human user to be a component of the distributed system integrated through unambiguous human–machine communication mechanisms offering protection against identity attacks.
7. **Consistent Experience Across Contexts**
   The unifying identity metasystem must guarantee its users a simple, consistent experience while enabling separation of contexts through multiple operators and technologies.

The *Law of Identity* is posted and discussed at `www.identityblog.com`. The "need to know" principle is a prominent underlying principle in the laws. The first four laws are the most universal ones and with the least political bent. The ideas for universal identity system are somewhat contentious, but privacy-ware interworking between identity management systems is a real issue. We note that the laws are intended for human identities and that object/entity identifiers represent a somewhat different case. This applies in particular to laws 6 and 7. Also, we should consider fully public services and their referential identities. These identities will be different, but this is also outside the scope of the above laws.

1.4 Aspects of Privacy

We shall now present and explain a few privacy attributes and traits. The list is not exhaustive, but we hope it will be a good starting point. In Chapter 3, we will add a few additional definitions (Sec. 3.2). We also recommend the so-called "Anon terminology" report [211] by Pfitzmann and Hansen for further definitions and explanations concerning "... Anonymity, Unlinkability, Undetectability, Unobservability, Pseudonymity, and Identity Management".

1.4.1 Data Privacy

The *data privacy* property can be seen as an access restriction in the sense of "protection from intrusion and information gathering" [112]. That is, a data privacy requirement means that the visibility of the object shall be restricted. Thus, in terms of a file system this would mean that the file should have restrictions on the read rights. That is, only people or processes[4] with the corresponding read permission are allowed to see the object.

Similarly, for a data object (message) in transit one must ensure that only those entities with proper permissions are allowed to read the object off the channel. The nature of the communication channels differ substantially; some channels may be able to provide physical protection while other channels, like radio channels, are easily read by any entity in the channels proximity. Various encryption schemes may be used to protect the channels.

So how strict is the privacy requirement? The question is relevant since some of the protection schemes are costly to operate and may cause considerable inconvenience to the users. The question is often best answered in terms of how valuable it is for you to keep the object private and who do you realistically see as your "enemy". The more important it is to protect the data and the more powerful your enemies are the better protection you should use. On the other hand, if you only want to protect against opportunistic casual prying by generally non-interested and less powerful enemies then it may be permissible to go for a lighter and less expensive protection scheme. This is analogous to other security/safety trade-off that we do all the time. The main difference is that it is much harder to properly evaluate the risks and correspondingly to judge what constitutes an adequate protection level for data objects.

[4] In the literature, one tends to use the term entity to denote "people or processes".

To enforce the data privacy property one can use different techniques. A data file on a hard drive is dependent on the operating system/file system being able to restrict access to the object. The operating system/file system can protect the data file in several ways depending on how strict the privacy requirement is, ranging from simple file system attribute settings to sophisticated use of data encryption techniques.

For objects in transit over a channel, one should generally assume that the channel is insecure. That is, unless one takes explicit action to protect the channel against eavesdropping one should assume that the channel is open for all to read. This also means that if data privacy is a requirement then one must explicitly protect the data when the data is in transit. The requirement applies to all "open" communication channels, which in principle includes all radio channels and all IP-based channels. The protection in question is almost always achieved through use of data encryption techniques to provide the security service *Data Confidentiality*.

1.4.2 Identity- and Location Privacy

Identity- and location privacy are independent, but nevertheless related, concepts. Again, the privacy property is about restricting access to information.

The *identity privacy* concept encompasses the right not to reveal your identity. *Identity privacy* is concerned with restricting access such that only authorized parties will be permitted to know the identity. Of course, an entity may have several recognized identities and additionally there are other references and addresses that may also serve to identify the entity. The *identity privacy* issue is particularly acute for wireless communication or communication over open channels like the internet.

Location privacy is the right not to reveal the location of the entity. Note that the "location" concept is not restricted to a physical location; it may well be a logical location[5]. The *location privacy* property is not too relevant if the location never changes. For instance, a stationary object does not benefit nearly as much from location privacy as does a mobile subscriber. However, as is discussed in the article by Øverlier and Syverson [205] even stationary entities (servers) may require a certain level of "location" privacy.

The location privacy property is associated with identity privacy in the sense that to know the position of an identified entity may have much more value to an intruder than just to know that there is an unidentified entity at

---

[5] This could be expressed as a relative location with respect to other entities or as an address/reference in a database etc.

the same position. The article "Subscriber Privacy in Cellular Systems" [172] takes a closer look at the current addressing model in cellular system and the associated identity- and location privacy issues. The article also discusses new solutions to the privacy problems in the wireless context.

1.4.3 Traffic Privacy

Traffic privacy is privacy aspects concerned with the pattern of exchanged data. This privacy threat is not concerned with data eavesdropping and, therefore, it can still be a threat for data confidentiality protected communication. That is, the data may be encrypted, but the traffic pattern will reveal information and that information may be sensitive. For instance, the frequency of an amount of transferred data may actually betray the type of data being transmitted. It may also be easy to link data transfers to known identities by analyzing the traffic pattern. So, traffic privacy threats may potentially serve to identify an entity, it may serve to obtain his/her position and it may serve to reveal what kind of activity he/she is engaged in. This again may be used as input in crypto-analysis and it may thus make it easier to break the data confidentiality protection too.

1.4.4 Movement- and Transaction Privacy

In the previous subsection, we briefly introduced the identity- and location privacy concepts. It is possible to take these concepts further and add a temporal dimension, i.e. to establish a time series of observations.

Movement privacy is your right not to divulge your movements. For a cellular subscriber this means that no one should be able to track your movements without your authorization. You may provide authorization (user consent) and the commercial so-called *location based services* (LBS) depends on you doing so. Movement privacy would also include data protection of your movement profile. This would include the property that GPS systems installed in cars do not leak information about the routes that you have traveled. It is noted that the movement privacy concept is not necessarily limited to identified entities. For instance, in military intelligence it may be of value to track the movements of entities even if their identities are unknown. Movement privacy is sometimes also called "untraceability".

Transaction privacy is your right not to divulge your transactions. The concept is broad and could include any transaction that you would want to keep private. For instance, you may not want to divulge that you made a call to a certain person at a certain time. Or you may not want your spouse to know that you met a certain person at a specific place/time. Transaction privacy is related to identity privacy and to location/movement privacy in the sense that if your movements are known then a lot of transaction information may be inferred or deduced. This is particularly true if one is able to correlate the information with information on the whereabouts of other entities or notable events etc.

1.4.5 Management of Privacy Sensitive Data

We briefly present the subject of management of privacy sensitive data here. In the subsequent section on *Privacy by Design* (Sec.1.5), we continue the investigation on management of privacy sensitive data.

1.4.5.1 Informed Consent and Fair Usage

Collection and use of private data should always require informed consent from the human user. We note that "informed" here requires that the respondent understands the ramifications of the consent. The data collected should be kept to a necessary minimum and the use of the data should be in accordance with the given terms.

1.4.5.2 Kept up-to-date

Management of privacy sensitive data is an important issue. Much of the data have limited lifetime and are subject to change. Information like your medical history is not static and information about your marital status may change and become outdated. It is not irrelevant that your medical history data is outdated, while other outdated data may still serve the intended purpose.

1.4.5.3 Correctness, Completeness and Consistency

You obviously should have the right to ensure that the stored information is correct and accurate. The authors of this book have surnames that frequently get misspelled. It is telling how difficult it can be to correct this information. The initial data entry may be a breeze, but corrections are often hard to achieve and there are often bureaucratic obstacles to be overcome.

Completeness and consistency is another matter. Quite often you do not want an external party to have a complete view of your private data, but it may matter that whatever data they have is consistent (for the given purpose). Consider that you previously may have had trouble with your mortgage. Then it is important that your new bank also knows whether you continued to default your mortgage or if you sorted it out. So completeness may matter quite a lot.

1.4.5.4 Ownership and Control

Then there is the issue of ownership. Who owns private data concerning you? The ownership issue concerns the right to copy and distribute the data and it concerns the right to erase the data.

For instance, if you decide to become the member of an organization, then you may have to give up some personal data to the organization. Contact data (street address, phone number, email address etc) is almost certain to be included in the data you must submit. It is not uncommon that they ask you for the right to sell the contact information to other companies/organizations. However, if you agree, it is very hard for you to later ensure that the contact copies are kept updated and used according to the agreement.

And what happens when you decide to quit your membership (or subscription or whatever). Under most jurisdictions, the organization/company is allowed to keep the data for a certain period of time. Many, if not most, organizations and companies will no doubt honor your privacy rights, but many don't really have the proper data management in place and so you risk that the data is never erased. One also needs to consider what would happen if they go bankrupt or are threatened by bankruptcy? Chances are that your data is sold off like any other asset. And, obviously, there are those organizations and companies that completely disregard your privacy rights.

1.4.5.5 Credible Protection

Honest intentions to protect your data is a prerequisite, but not in itself sufficient. The collected private data need to be properly protected during the entire lifetime of the data item, and then it should be discarded in a proper manner. In a digital setting this inevitably means using encryption, but we may add that this has to be done properly and there has to be a supporting architecture. That is, there needs to be a security architecture in place and it has to be designed to protect private data. It is of limited use to have strong encryption if proper management of the secret keys is missing. This would include cases where too many people have legitimate access to the data without having a genuine need to the said data.

1.5 Privacy by Design

1.5.1 Privacy as a Mandatory Feature

There are some services that are essential and cannot be dispensed with. Consider a large scale system like the telecommunications network or a more contained system like your car. Safety and security are must-have services and they are fundamental in the sense that systems are required to have some minimum level of services implemented. To some extent the right to these services is so fundamental that you don't even have to require them when purchasing services, and furthermore they might even be considered irrevocable rights. That is, in the extreme, national laws may explicitly forbid contracts that revoke these rights. Privacy aspires to the same fundamental service level as does safety and security, and indeed these do have intersections where they also affect each other.

However, in order to have fundamental services in place one cannot add them as afterthought after the product or service has been produced. "Safety by design" is a well-known concept in industry and likewise "security by design" has emerged as a concept in the software industry. "Privacy by design" is not quite with us yet, but the importance of built-in privacy features has clearly been recognized and is now a primary requirement in EU research and development programs. Likewise, we have privacy commissioners in many countries and privacy officers in many companies, and gradually we expect that privacy will be truly recognized as a core service.

1.5.2 The 7 Foundational Principles

The *Privacy by Design (PbD)* challenge is an initiative from Ontario, Canada. It may also be stated as *Privacy by Default*, which is essentially what it represents. In the following, we present the so-called "7 Foundational Principles" that have emerged from the *Privacy by Design* initiative [200].

1. Proactive not Reactive; Preventative not Remedial.
2. Privacy as the Default Setting.
3. Privacy Embedded into Design.
4. Full Functionality; Positive-Sum, not Zero-Sum.
5. End-to-End Security; Full Lifecycle Protection.
6. Visibility and Transparency.
7. Respect for User Privacy; Keep it User-Centric.

The principles have received broad endorsement, but will still be seen by some as political and slightly activist by nature. We shall not discuss the politics of privacy, but rather present the principles more-or-less as-is. The presentation below is an adapted and abridged version of the principles as stated in [200].

1.5.2.1 Proactive not Reactive; Preventative not Remedial

The Privacy by Design approach aims at anticipating and preventing privacy invasive events before they happen. That is, one should not wait for privacy risks to materialize or only offer remedies for resolving privacy infractions once they have occurred. It is seen as far better and safer to prevent privacy incidents from occurring in the first place.

1.5.2.2 Privacy as the Default Setting

Privacy should be the norm and hence it should by default be set to turned on. This will ensure that that personal data are automatically protected in any given IT system or business practice. Thus, from a user perspective, doing nothing with respect to configuration should result in their privacy being maintained at an acceptable level. Jokingly, the acronym PbD is sometimes also referred to as Privacy-by-Default.

1.5.2.3 Privacy Embedded into Design

Privacy must be embedded into the design and architecture of IT systems and business practices. Privacy must be designed to be an integral part of the system, without diminishing functionality. Policies for operations and business practices must adhere to privacy principles as a core component of the system. This also applies to emergency cases.

1.5.2.4 Full Functionality; Positive-Sum

Privacy by Design seeks to accommodate all legitimate interests and objectives in a"win-win" manner. Research has shown that people's perception of the trustworthiness of a system increases when the system clearly demonstrates its intention and ability to protect itself. It is contended that a system demonstrating ability and willingness to comply with privacy requirements will be trusted by the users, and that in fact user will over time expect and require systems to be trustworthy with respect to privacy. Possible conflicts, such as privacy vs. security, are seen as requirements- and design shortcoming rather than inherent and irreconcilable conflicts. In fact, the PbD approach actively encourages both strong security and strong privacy.

1.5.2.5 End-to-End Security; Full Lifecycle Protection

A prerequisite for Privacy by Design is to have strong security measures in place. Indeed, strong security with end-to-end capabilities are essential to privacy and without strong security there can be no privacy. Full lifecycle protection also means that security and privacy are very much part of the evolution of the system and that one considers the lifecycle of private data. Thus, sensitive data must be kept secure and private from the moment the data is collected by the system until the data is securely destroyed or declassified. There should be no gaps in either protection or accountability.

1.5.2.6 Visibility and Transparency

Trust is an essential component of human interaction, but trust is not enough. In a sense, trust is about the intentions and while one certainly wants assurance about the good intentions of someone or something, one also need assurance about the corresponding party's ability to keep promises and act according to stated intentions. Thus, there is a distinction between trust and trustworthiness, and this distinction is also present for privacy trust.

To ensure privacy trustworthiness, one needs to have visibility (of intentions) and transparency (of actions). Proper visibility and transparency will enable accountability, which is very much a needed attribute for trustworthiness. Another aspect is compliance to stated privacy policies. Privacy trustworthiness will also require visible and transparent verification of claimed compliance. Necessary steps to monitor, evaluate, and verify compliance with privacy policies and procedures is therefore essential.

1.5.2.7 Respect for User Privacy; Keep it User-Centric

A user-centric approach means that individual users, who have the greatest vested interest in the management of their own personal data, must be able to retain control over the privacy sensitive data.

The privacy sensitivity is subjective to the human user, and it therefore stands to reason that the user has the last word on management of the data. Obviously, there are other rights than privacy rights that must also be respected and addressed, but, in general, privacy sensitive data should under control of the user. Respect for user privacy extends to the need for human–machine interfaces to be human-centered, user-centric and user-friendly so that informed privacy decisions may be reliably exercised. Similarly, business operations and physical architectures should also demonstrate the same degree of consideration for the individual, who should feature prominently at the centre of operations involving collections of personal data.

1.6 Summary

When discussing privacy it is important to be clear on what is being discussed. It is also important not to get too emotional, after all, threats on privacy tend to feel personal. Much, if not most, of the privacy sensitive data is not critically sensitive, and the corresponding privacy loss is limited. Data mining, of course, may turn this upside down and re-combing and correlate seemingly innocent data such that too much is revealed.

Most, if in practice not all, internet attacks on your privacy will be impersonal. It may not feel impersonal, but in reality there are few specific targeted attacks out there. That is not to say that privacy attacks will not occur – they most certainly will, but we expect most of these to be annoyances and in fact be legitimate in the sense that users have authorized or facilitated the data collection in the first place. Suffice to mention so-called social networking and the amount of private data that people willingly put out on Facebook, Google+, Twitter and similar sites.

In the paper "Where do all the attacks go" [110], the authors ponder the fact that there appears to be an "enormous gap between potential and actual harm" when considering all the attacks reported on the internet and when considering the proliferation and abundance of powerful malware present on the internet. The authors conclude that it in the grand scale of things it just isn't economically feasible to attack single individuals and that diversity, in terms of software (incl. software versions) and hardware used, makes it very hard to attack people. In an unrelated paper, "Gaming security by obscurity" [209], the author used gaming theory and found that security by obscurity is indeed working in a defense-in-depth kind of way. This further underscores the configuration problem the attacker has in deducing the state of software and hardware in the target computers and lends credibility to the conclusions that broad large-scale attacks are indeed difficult to carry out. Not that this will be of any comfort whatsoever to those that suffer an effective attack. And, of course, privacy is a little different than security here in that retrospective attacks on collected and stored data could be valuable, while retrospective attacks on security often tend to only be of historic interest.

Privacy trust and privacy trustworthiness is becoming more important over time. Businesses that can demonstrate privacy trustworthiness are likely to do well, and this stems not only from being able to handle privacy but also for demonstrating an ability to care for the customers.

In summary, we advocate prudence in terms of data privacy and we advocate user awareness. For the professional designing and implementing systems we very much recommend being clear on the security and privacy goals and requirements. We also recommend keeping a *Privacy by Design* mindset when designing and developing new systems.

2

The Legal Context of Privacy

In this chapter, we examine Emergency Call (EC) services, Lawful Interception (LI), Digital Forensics (DF), and the Data Retention Directive (DRD). These have all the potential to intrude on our privacy, but have a legal justification and aim at protecting society at large. For the emergency call service, the justification is to help you in case of distress and emergency. We also briefly look at Digital Rights Management (DRM), where intellectual property protection sometimes is at odds with privacy. We round off with a brief look at some surveillance technologies. The legal aspect here is that the right to protection, safety and security is trumping the right to privacy.

2.1 Emergency Call Services

2.1.1 Emergency Call Requirement

Telecommunication regulations require that emergency handling authorities are given access to all available data concerning the emergency call. So if you call 112 (within EU), 911 (U.S.) or some other dedicated emergency call number, the EC operator will get access to all information about the call that the telecom operator has available. This will override any restriction on presentation of the A number (the caller number) and it will provide the EC operator with the corresponding physical address. For a mobile subscriber, this is not the subscription address (billing address), but rather the geographical coordinates of the caller at the time of making the call. That is, to the extent that the cellular operator is able to determine the location. There are requirements on the operator in this respect, and so the operator has a legal obligation to be able to determine the approximate position of the caller. The emergency call handler function is formally known as a Public Safety Answering Point (PSAP). The ETSI Special Report on emergency call handling [93] is a useful reference for emergency call requirements.

2.1.2 U.S. Regulations

In the U.S., the FCC has a basic 911 rule that requires wireless service providers to connect all emergency calls to a PSAP. The requirement is strict, and it applies also when there is no valid subscription (applies in particular to cellular phones). Phase I of the Enhanced 911 (E911) rules require wireless service operators to provide the PSAP with the telephone number of the originator of the call and the location of the cell site or base station transmitting the call. Phase II E911 of the rules require wireless service providers to provide precise coordinates of the caller, accurate to within 50 to 300 meters (depending upon the technology used). The E911 standard is not so strict for indoor calls due to occasional technical difficulties in determining accurate position and due to the fact that the caller is likely to be able to provide the information himself/herself. As of 2011, the FCC requires the operators to explicitly list areas where they cannot comply and to provide justification for the non-compliance.

2.1.3 EU Regulations

The EU requirements dictate that the calling line number, and indication of the emergency caller's position be made available to the PSAP function (See Chapter 4.2 in [93]). Like in the U.S. it applies even when there is no valid subscription. In many countries, this is interpreted even to apply for phones without a subscription module (SIM card or UICC card). The serving network would normally have all the required information, but there are exceptions for IP-based services. The exceptions are mainly concerned with establishment of the subscriber's identity. The EU regulations also explicitly mention multi-media call, email communications and messaging (SMS/MMS), but there are no explicit requirements.

Within the EU, the European Telecommunications Standards Institute (ETSI) has a special responsibility for preparing telecommunications standards. The Emergency Telecommunications (EMTEL) work is carried out in many different working groups, but the work is supervised by a special committee (SC EMTEL). More on ETSI EMTEL at `www.emtel.etsi.org/`. The ETSI EMTEL technical report TR 102 229 [97] provides an overview over regulatory texts concerning emergency call handling, for the EU, and TR 102 180 [96] provides a generic technical report on the topic. Figure 2.1 is transposed from figure A.1 in [96] and depicts the basic EMTEL architecture.

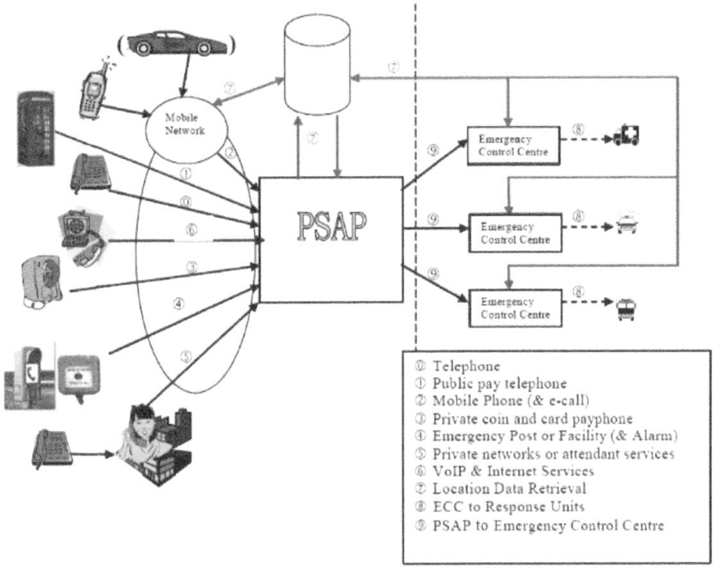

Figure 2.1 Emergency Communication – Functional Architecture

2.2 Lawful Interception

2.2.1 The Justification for Lawful Interception

Lawful Interception (LI) is a means for a Law Enforcement Agency (LEA) to obtain legally sanctioned access to private communications, such as telephone calls or e-mail messages. LI is, by its very nature, privacy invading. However, LI is a specific measure that applies to individuals or groups that are suspected to be involved with criminal activity. As such it is an individually targeted measure, and LI is of course only intended to be granted by a court or other jurisdictional entities. Specifically, LI rights are not assumed to be granted by the LEA itself. LI, as a measure against organized crime and terrorist activities, is largely seen as a just measure. The threats against society from organized crime and terrorist activity are real and in the post 9/11 era most people will agree that law enforcement agencies must be given proper tools to protect society against serious criminal activities.

2.2.2 Foundation for Lawful Interception

Modern LI has its foundation in the European Council Resolution of January 1995 [99]. There, the "International Requirements for the Lawful Interception of Telecommunications" (IUR) was outlined. This was the result of several years of work by the European governments in cooperation with Australia, New Zealand, Canada and the USA. Within the EU, it is ETSI which has the responsibility for developing LI standards, but ratification and transposing of the standards remain a national responsibility. The LI standards and technical documents published by ETSI covers the whole spectrum of interception aspects, from a logical overview of the entire architecture and the generic intercepted data flow, to service-specific details for e-mail and Internet.

2.2.3 Telecom LI Standards

ETSI has published quite a few LI specifications, including specifications for the (cellular) handover procedure [95, 98]. These specifications map out the dataflow for the intercepted data in the telecommunication networks or services. They specify the network and service protocols necessary to provide LI, as well as the physical or logical interception points both for packet data and switched-circuit communications (telephony mostly). Other relevant specifications cover LI network functions [92]. A generic LI network architecture is outlined in the technical report "Concepts of Interception in a Generic Network Architecture" [94].

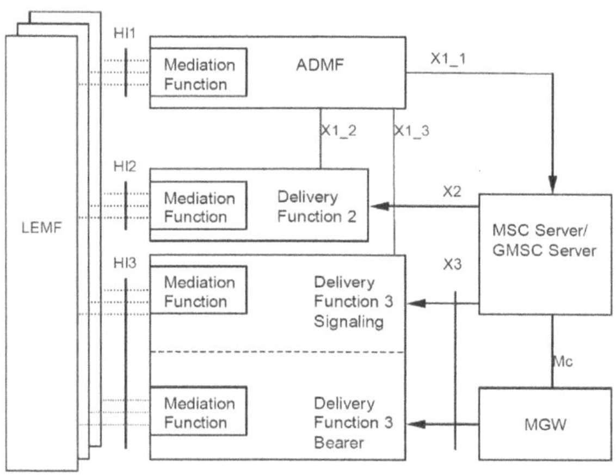

Figure 2.2 Lawful Interception of circuit switched call in the 3GPP architecture

The specifications are subject to constant review and updating within ETSI to accommodate emerging needs. The trend towards all-IP networks has necessitated new standards for IP-based interception, captured in the TS 102

232-x series. The article "Lawful Interception" [239] provides an overview over European LI standards. The above standards are generic and have to be mapped onto the specific systems. Figure 2.2 depicts the "circuit switched intercept configuration" as defined in 3GPP TS 33.107 [12].

2.2.4 The Dark Side of Lawful Interceptions

2.2.4.1 Subverted LI

LI systems may of course be abused. An example of this is the high-profile LI abuse case from Greece. Basically, the intruders installed rogue software that subverted its built-in wiretapping features for their own purposes. The intrusion and LI-abuse lasted from August 2004 until January 2005 [213]. The intrusion was very professional and included interception of the Greek prime minister's phone amongst others. Technically, "The Athens Affair" intrusion was very advanced and consisted of rogue software being installed on Ericsson servers. The software then permitted illicit use of the mandatory-to-install LI system. The "Athens Affair" should serve as a warning to the risks of wilful introduction of security holes. Irrespective of the justification for lawful interception there are serious systemic problems with the introduction of "security holes" in a system architecture.

2.2.4.2 Just Laws

Within the telecom standardizations world, LI systems are considered mandatory features. So, LI systems are implemented and deployed. One may or may not feel compelled to trust the telecom operators or the telecom vendor, but barring isolated incidents the risk of illicit lawful interception is fairly low. It will never be zero, but unless you are a very important, rich or famous person, chances are that LI intrusion is less likely than other means of illicit surveillance and eavesdropping.

However, the telecom operators and vendors do not decide what "lawful" is supposed to mean. This is the realm of the national state and their legal system. So, the LI abilities will be available to benevolent states as well as to dictatorship states with severe records for oppressing its population alike.

Another matter is that the law enforcement measures should be justified in context of the suspected crime. Misdemeanor type of offences should clearly not justify the use of LI systems. On the other hand, a murder investigation may easily qualify for LI usage. The tricky question is to decide exactly where the line should be drawn on this matter. There is no universal answer here, and practices vary considerably.

2.3 Digital Forensics

2.3.1 Digital Evidence

In all the criminal investigations, the law enforcement agency must obviously scrutinize the crime scene thoroughly. And, increasingly, the crime scene includes computers or other digital devices. So when the media report on investigations into financial fraud, suspected terrorism, and organized crime, they frequently mention the importance of digital forensic evidence. The forensic evidence in question may be files found on computers, but even when the suspected criminal has attempted to delete the incriminating evidence it is quite likely that the system would still contain meta data and large amounts of "undo" feature data.

As reported in the article "Computer Forensics" [28] a computer can be used in a variety of crimes and a computer will likely also contain data and meta data associated with the transactions. The problems the digital forensic scientist faces are similar in some respect to that of the privacy intruder; he or she must uncover hidden data and restore deleted files etc. However, in digital forensics the gathered evidence is supposed be valid in a court of justice, and then the requirements on how the evidence is obtained is strict [28].

2.3.2 Clouds on the Horizon

Digital forensics, while being a powerful investigation tool, is not a perfect tool. Increasingly, we find that identity schemes are becoming more indirect and that data are not always stored locally. Indeed, with cloud computing looming large it is increasingly likely that much, if not all, the data is stored somewhere in "the cloud". This doesn't mean that the data cannot be investigated and mined, but it does mean that an able criminal may add complexity to the process. The complexity does not only stem from technical difficulties, but also from legal difficulties. Imagine for instance that data concerning a crime in country **A** is store in the cloud in country **B** and **C**. This, seen from the criminals point of view, is a measure of defense in depth.

Another aspect is that the methods that were used to enhance the privacy of innocent individuals can also be used to protect criminals. So, it is no surprise that criminals have gotten better at data privacy (disk encryption and protected communications). This, of course, represents real problems for digital forensics. Garfinkel outlines some of the challenges for digital forensics in the years to come in [117].

We shall have more to say on cloud computing and privacy in Chapter 12.

2.4 The Data Retention Directive and Similar Measures

2.4.1 Introduction to the Data Retention Directive

The Data Retention Directive (DRD) [102] is an EU directive that permits lawful surveillance of telecom traffic. In contrast to LI, the data retention requirements applies, in principle, only to control plane data. That is, it does not apply to the data content. Another difference is that DRD applies to all user traffic and to all user connections. That is, data is captured and stored without any *a priori* suspicion against the users.

Within the EU regulatory regime the requirements will be to store call related data for 6 to 24 months, which is far in excess of the charging related requirements. The motivation for the Data Retention Directive is prevention of crime- and terrorist acts. The DRD, as defined here is an EU specific directive, but similar schemes are in place (or about to come in place) in non-EU jurisdictions. In the article "The EU Directive on Data Retention – An End to Justify the Means" [230], the former CTO of Telenor Group provides an overview of the DRD as seen from a telecom operator's perspective.

2.4.2 Historical Background

The 9/11 terrorist attacks in 2001 created a new political environment, but the impact in Europe was limited until the Madrid bombings in 2004. The Madrid bombings on March 11 2004, killing 191 people and wounding around 1800, directly triggered a meeting of EU Member States that led to the publication of the European Council's Declaration on Combating Terrorism [100]. This document was the starting point for the subsequent proposals for rules relating to the retention of communications traffic data by service providers. Following discussions of the proposals and the legislative process, a new Directive on Data Retention was finally adopted by the EU in February 2006.

2.4.3 Privacy vs. the DRD

The EU had a privacy directive [101] in effect prior to the DRD. The privacy directive contains harmonized (transnational) rules relating to the protection of personal data whenever traffic data are processed in relation to electronic communications services. The privacy directive generally states that traffic data should be deleted or made anonymous when they are no longer needed for the effectuation of the communication, or for charging purposes. But access to traffic data is of course important to ensure proper identification of subscribers and users of services whenever this is needed for purposes relating to law enforcement and security. Thus, the privacy directive opens for softening the restrictions relating to data protection whenever it is required for national security, public security and/or prevention of crime. Privacy requirements may thus be overruled if it is deemed necessary to combat crime and terrorism.

2.4.4 DRD Requirements

The DRD applies to all the providers of publicly available electronic communications services and public communications networks. This covers fixed-line and mobile telecommunications operators, satellite operators, cable operators, Internet service providers and companies that provide electronic communications services such as web mail, instant messaging etc.

The data to be retained include information both on successful calls and unsuccessful call attempts. It should permit tracing and identification of the source and destination of a communication, the date, time and duration of communication, the type of communication and the equipment used for the communication and its location. For fixed-line telephony, retained data will include the telephone numbers of callers and those receiving the calls, but also numbers involved in rerouting, names and addresses of subscribers or registered users, and the type of telephone service used. For internet services, data revealing users' identities (ID) and IP addresses must be retained. Data related to mobile services must also include the international mobile subscriber identity of callers and those receiving the calls and the international mobile equipment identity of both parties. Thus, for cellular system one not only collects the telephone numbers, but also the identity of the subscriber, the identity number of the mobile device and the location data. The cellular identifiers and cellular privacy aspects are covered in depth in Chapters 6–8.

All categories of data covered by the directive are required to be retained for a minimum of six months and a maximum of two years. Member States are free to set the retention period within these limits.

2.5 Digital Rights and Digital Wrongs

2.5.1 Intellectual Property Rights

With digital content (music, pictures, films etc) comes the issues of digital copyrights. The fact that the authors or content owners have intellectual property rights to the material is not generally disputed. The problem with digital rights is how to enforce the rights. The problem stems from the fact that it is essentially without cost to copy and distribute the fully digitalized material. This makes it very easy for people to make perfect copies and distribute the copies to their friends. The problem is exacerbated by the fact that making a limited number of copies may be perfectly legal. The copies may be permitted for use on different playing devices (by the same person) and for having a backup copy (in case of disk crash etc). It is not easy to enforce digital copyrights without either needlessly restricting the usage of the digital object or by being too intrusive in the enforcement.

2.5.2 Digital Rights Management and Privacy Violations

Digital Rights Management (DRM) is used to denote systems that enforce intellectual property rights. So far so good. However, many people consider DRM systems to infringe on their legal user rights. Adding insult to injury, DRM systems are also commonly considered to be user unfriendly.

2.5.2.1 The Sony DRM rootkit

As with lawful interception and the "Athens Affair", we have cases where intellectual property right enforcement have gone bad. One particularly noteworthy case is the so-called "Sony DRM Rootkit" scandal [224]. The scandalous DRM software, which was first uncovered by Mark Russinovich [219], was aimed at protecting music CDs. Sony, without informing the users, included the Extended Copy Protection (XCP) and MediaMax CD-3 software on some of their music CDs. This software was automatically installed on Windows computers when customers tried to play the CDs. The software not only installed itself without user consent, but it also hid itself in a manner reminiscent of a so-called rootkit. A rootkit is software that enables privileged access to a computer while actively hiding its presence from the users and administrators by subverting standard operating system functionality or other applications. The term "rootkit" has strong negative connotations through its association with malware (malicious software). In fact, the Sony DRM software was initially detected by specialized rootkit detection software.

Sony BMG received a lot of negative attention due to the DRM software and had to withdraw the DRM software amidst a storm of legal trouble. Ironically, the software was also found guilty of infringing on intellectual copy-rights by including open source MP3 encoder library without acknowledging it. An interesting side of the Sony DRM Rootkit scandal is that many of the Anti-Virus (AV) vendors were late, to say the least, in detecting and reacting to the rootkit software.

2.5.2.2 Genuine Advantage Anyone?

There have been many incidents where DRM schemes have reported a lot more back home than they really need to, and this is certainly a threat to your privacy. It could be relatively innocent like the 2006 controversy regarding the Microsoft Windows Genuine Advantage (WGA) phone home functionality [199, 216]. The WGA software, which was intended to combat illegal copying of the Windows operating system, called home to the Microsoft servers every day. The problem here is that the legitimate paying customer is adversely affected by this software and that furthermore Microsoft didn't tell you that the software would be phoning home. And how would we, the law abiding citizens, really know if the WGA tool does not spy on us?

2.6 Summary

The justification for LI is altogether a solid one, and it is a targeted and subscriber specific surveillance service. The DRD and similar schemes, on the other hand, are much more broad in scope and aim at the whole subscriber population. This is a "guilty till proven guilty" approach. The amount of data that is being collected varies, but generally what is collected is control data and not the conversation as such. Still, this is a massive invasion of people's privacy and it takes place irrespective of whether you are a suspected criminal/terrorist or not. The DRD is surely a draconian measure, but there is nevertheless quite a lot of popular support for these types of measures. The argument goes along the lines of "if you are innocent then you have nothing to fear".

This is all fine, but again there are powerful counter arguments along the lines that innocent people have a right to privacy. The famous Franklin quote [112] illustrates that this problem has existed for some time. There is a real fear that powerful tools as the DRD, used with data mining tools and in combination with other databases etc, permits scenarios reminiscent of Orwell's 1984 [8] totalitarian surveillance state. So the question remains; should we trust Big Brother? And, realistically, what is the alternative. How much privacy must we be ready to give up to obtain a minimum level of safety? The jury is still out, and, of course, there is no definitive answer, just difficult questions.

Part II

Privacy in the Real-World

3

Anonymous Communications

3.1 A High-Level View of MIXes and Onion Routing

Digital MIX networks and Onion Routing (OR) are related concepts, but they are fairly different in their intended purpose. Both support "anonymous" communications, but we need to be clear on what we mean by "anonymous" communication to see the differences. This chapter will not go into deep technical explanations concerning the implementations and theories about specific constructs, but will rather focus on the *services* that MIXes and Onion Routing offer in general. So, the presentation is fairly high-level and is typically concerned with questions such as what the properties are, what the limitations are and what these systems are really good for. We will provide a set of references for the interested reader to dive into the details, amongst then a few survey/overview papers, including [72, 73, 222].

3.2 Privacy in Context

3.2.1 Privacy Cases

To illustrate the functional aspects of MIX and OR, we outline some examples of different privacy aspects. First, we define our principal entities.

- $Alice$ is one or the communications partners
- Bob is the other principal communications partner
- Eve is the privacy intruder/adversary
- $Node$ is a network node (incl. an OR router or MIX)

We could have defined other parties too, like*a)* an opportunistic external party that will eavesdrop whenever possible but who otherwise will be well behaved and *b)* like a government agency with lawful interception authorization etc. Likewise, one sometimes differentiates between different types of intruders (passive vs. active) etc., but we do not need these to illustrate the concepts. Note that Eve may attempt to masquerade as any other party, or she may attempt to trick, intimidate or compromise other parties to get information. Eve may even be a principal participant, let's say Alice's evil twin sister Malice. In the literature, there is often a sharp distinction between the principal parties and the intruder/adversary, but in the real world a dishonest principal is quite possible. The logical extension to the Node will be a set of nodes, and ultimately the set could encompass the entire network.

3.2.1.1 Case A: Alice wants to remain anonymous

In this case, we have that Alice wants to communicate with Bob, but Alice does not want to reveal her identity to Bob (or Eve). Alice does not trust the Nodes too much either and would strongly prefer that they too remain ignorant about her identity. This scenario is roughly analogous to Alice sending a letter through the Post Office system without including her name or address. However, we do want Bob to be able to answer Alice.

3.2.1.2 Case B: Alice and Bob want to communicate privately

In this case, we have that Alice and Bob want to communicate privately, but they do not need (or want) to remain anonymous with each other. However, they do not want Eve to learn that they are communicating and they would prefer to avoid the Nodes from knowing. This scenario is different from the Post Office analogy in that the envelope should not have Alice' or Bob's address on it.

3.2.2 Privacy Concerns and Properties

The common theme to **Case A** and **Case B** is that it is primarily the address/identifier information that causes concern. Since we are primarily dealing with IP packets, this will translate into the "Source IP address" and the "Destination IP address", but other fields may also convey "traceable" information. We note that it will be insufficient to use IPsec [163] to encrypt the payload as it is the address information that worries us. One may use ESP in tunnel-mode (Fig.3.1), but it will not completely solve the problem since there needs to be IP addresses in the outer header too. ESP in tunnel-mode is very useful for two domains (sub-nets) to communicate securely over intermediate networks. The domains route all external traffic via security gateways and the addresses in the outer IP headers will then belong to the security gateways.

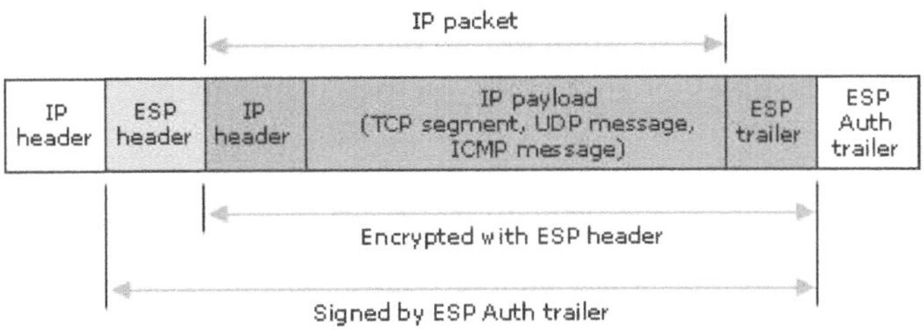

Figure 3.1 IPsec w/ESP in tunnel-mode is not enough

3.2.2.1 Anonymity vs. Pseudonymity

By *anonymity* we shall mean that one cannot distinguish the entity/subject and identify it/him/her. Anonymity is relational and one (Alice) may be anonymous with respect to some parties (Eve) while at the same time be well-known to other parties (Bob). A temporary random identifier may be used as an alias identity, and this will allow Alice to have an anonymous referential identifier. The trouble is that prolonged use of the referential identifier will mean that it emerges as a distinguishable identity. Still, with care, Alice could use the referential identity without her communications parties being aware of her "true" identity.

An anonymous referential identifier will, over time, be transformed into a pseudonym or an alias. The concept of *pseudonymity* can, therefore, be defined as using a pseudonym as the identifier.

Observe also that prolonged use of a pseudonymous identifier will inevitably lead to information leakage [71, 133] in the pattern of communication, network latency etc., and that the pseudonymous identity will potentially become so well-known that other parties will be able to infer the association between the pseudonym and the "true" identity. The association may be betrayed on many levels as demonstrated in [250] where group memberships in social networks allow the identity to be inferred. The user (Alice) will probably not be aware that she has lost her anonymity/pseudonymity.

Identity privacy is not the only aspect one needs to assess. Anonymity may be fine for some cases, but very often there is a credible requirement for accountability. And a truly anonymous entity cannot be made accountable. Alice may need to hide her identity from Bob (and external parties too), but Bob may have a genuine requirement for accountability. Pseudonymity can sometimes be combined with accountability and this may allow us to have it both ways. There are a number of mathematical models on how to measure and classify anonymity. We shall not go into any detail here, but instead refer to [162] where a survey of anonymity metrics is provided.

3.2.2.2 Unlinkability

The concept of unlinkability is also important to consider. In the Alice and Bob cases, unlinkability is the characteristic that, as seen from an external party, there should be no apparent relationship between the parties after the communication has ended. We distinguish between unlinkability between objects and between parties:

- Unlinkability of objects (messages, events, etc.)
  As seen from the intruder, the objects shall appear to be no more or less related after the transaction took place.
- Unlinkability of parties (relationship anonymity)
  As seen from the intruder, the sender and receiver shall appear to be no more or less related after the transaction took place.

Unlinkability is a sufficient condition for anonymity, but it is not a necessary condition for anonymity [211].

3.2.2.3 Unobservability

Unobservability is another possible goal for the communications partners. It is somewhat different in the sense that this goal is about external parties not being aware that there is communication at all. The signal is masked out in noise or it is carried out on a channel that the adversary has no access to and no awareness of. In some ways, unobservability is not a privacy goal, but on the other hand, if unobservability is attained then the adversary will not even know that there is any communication. Traffic analysis may benefit from the fact that the adversary is aware of the communications intensity and distribution. For instance, it is a well known historic fact that major military maneuvers are often proceeded with a marked increase in the communications intensity. The opposing side signals intelligence (SIGINT) unit would do well to warn about the event since it may be likely that there would be an attack. Of course, one may introduce noise (dummy traffic) to mask genuine communications, and this is a well know counter tactic.

Another strategy for hiding the fact that there is "meaningful" communications is to hide it inside (apparently) non-meaningful information. The word "steganography" is of Greek origin and means "concealed writing". Steganography is basically a way of hiding messages in such way that no one, apart from the sender and intended recipient(s), even suspects the existence of the message. There are many ways to do this. A popular example is to take the least significant bit(s) of a high-resolution picture and use these bits for conveying information. The picture would still look very much the same to the casual viewer. The message would literarily be "hidden in plain sight". The parties should also encrypt the hidden message just in case it was detected. There is terrible overhead in steganography, but this may not matter for the intended application. Below is a steganography example. The picture of the cat, Fig. 3.3, is hidden within the picture of the trees, Fig.3.2.

Figure 3.2 Steganographic host picture (source: Wikimedia Commons)

"Image of a tree with a steganographically hidden image. The hidden image is revealed by removing all but the two least significant bits of each color component and a subsequent normalization. The hidden image is shown below." - *Source: Wikipedia*

Figure 3.3 Steganographically hidden cat (source: Wikimedia Commons)

Note that the original pictures have been converted to black and white. A good starting point for further reading in Steganography would be [59, 184].

3.2.3 Privacy Assets and Privacy Goals

In the above, we had two different cases where we saw that Alice and Bob had different views on what to protect. And, this is important: We need to clearly define what the privacy assets are.

In Case A, we had that Bob didn't have identity privacy as a goal. For Alice, her identity is a privacy asset (in both cases) and the value of the asset will be diminished if Bob (Case A) is allowed to learn her identity. In case B, she may allow Bob to know her identity, but it is still vital that Eve does not learn her identity We must consider whether the assets are vulnerable and/or exposed in any way. In itself exposure and vulnerability does not translate into a threat. To have a threat one also needs a threat agent (adversary).

In Case A, we may imagine that an intermediate node learns Alice's identity. The Node has no interest in the identity and so the Node purges the identity from internal memory almost immediately. Clearly, the identity was exposed and vulnerable, but unless there was some entity that actually would want to exploit the vulnerability then the exposure was harmless. Our adversary, Eve, wants to learn the identity of Alice, and she therefore poses a threat towards the asset. It may be useful to imagine that Alice's identity is an asset to Eve as well, but it just so happens that if Eve gains the asset then Alice will stand to lose. This could easily be a negative-sum game, where the gain Eve has is less than the loss Alice experiences. In fact, there is often an asymmetry in the privacy loss/gain utility. Alice must now try to decide how valuable the asset is to her. Does she really need full anonymity? Will pseudonymity suffice? How much time and energy will she spend to protect the asset? What are the threats and who may the adversaries be? How powerful are the adversaries and how well motivated and aggressive will they be in achieving their goals?

3.2.4 Privacy Invasion Intruders

3.2.4.1 The Dolev-Yao Intruder Model

The Dolev-Yao (DY) intruder model [81] is a simple but powerful model. The DY intruder is capable of intercepting all communication channels and it can manipulate messages at will. The DY intruder has also the capability of storing all data ever exchanged on the communication channels. The DY intruder is able to use its knowledge and abilities such that if an attack is theoretically possible then the DY intruder will execute it. The DY intruder will attack and intercept on live sessions, but it may also use recent information to go back in time and retrospectively break old protocol runs. This may allow the DY intruder to falsify historic protocol runs.

The DY intruder is clearly a very powerful adversary, but there are limits to the DY intruder powers. For instance, a DY intruder is not capable of physical intrusion of the principals. Another point is that while the DY intruder can manipulate messages at will, it cannot actually break the cryptographic primitives. It may still be able to break the crypto-system due to incorrect assumptions and wrong use of crypto primitives. There are variations of the DY intruder, including so-called wireless intruders and over-the-air intruders (see Sec.8.2.8). These are slightly less powerful, but in general it will be hard to find meaningful differences concerning the attacking abilities. There are also more powerful intruder models, including models in which the intruder can carry out physical intrusions etc. These may be more realistic in some sense, but real-world intruders also have a lot of limitations, and this isn't reflected in the DY-based models.

3.2.4.2 Realistic Intruder Models

The DY intruder and variants thereof is not the typical adversary. A classification of realistic would-be intruders and their respective budgets are shown in [42]. The classification is assumed to be largely valid today (see Chapter 4.4 in [91]), but one may assume that the potential number of instances of each class has increased considerably. Furthermore, cloud services and ready-made "script kid" attacks will allow a new economy of scale in the attacks.

The classification in Table 3.1 is amended slightly from the original [42]. In the original table, the budget was assumed to be used on FPGA and/or

Table 3.1 Intruder Resource Classification

Intruder:	Resources:	Comment:
Hacker	Desktop PC; $400 budget	Single person intruder
Small organization	Multiple PCs; $10.000 budget	Experts available
Medium organization	Many computers; $300K	Many experts available
Large organization	Many computers; $10M budget	Many experts available
Intelligence Agency	Many computers; $300M budget	Many experts available

ASIC hardware. Today, one is probably better off buying machines with large stacks of graphic processor (GPU) boards.

3.2.4.3 Privacy Intruder Capabilities

We have chosen not to define a specific "privacy intruder". We consider that "all information may be sensitive" and in this perspective a standard intruder will do just fine. The only real difference is that meta information (header information) must now explicitly be considered as targets for the intruder. We still want to classify some traits and types of intruders. These should be viewed in context with Table 3.1. The following bullet-point list should illustrate this.

- *Passive vs. Active*
  A passive intruder is only allowed to eavesdrop. The active intruder will trigger events to gain advantages.
- *Memory capability*
  Some theoretical intruder models will allow the intruder to store every piece of information ever retrieved. A real intruder will have limitations.
- *Reasoning capability*
  A DY intruder is able to use all information in an optimal way. Obviously, a practical intruder, however capable, will not achieve this.
- *Crypto-breaking capabilities*
  It has been customary in the literature to assume "effective crypto", i.e. it{ is assumed to work perfectly. This is a naive assumption on at least two major accounts. First, the key length or similar may be too limited and exhaustive breaking efforts may be feasible. Second, the crypto primitive may be flawed and analysis may "break" the primitive.
- *Message inserting*
  An active intruder may insert messages in the message stream.
- *Message deletion*
  An active intruder may delete messages in the message stream.
- *Message modification*
  An active intruder may modify messages in the message stream.
- *Spatial presence*
  If one wants to eavesdrop on a geographically distributed system, like a cellular system, then the eavesdropper may need be present in all locations. While no practical intruder can be omnipresent, a common way to model the intruder is to assume that the intruder is the network. That is, everything will be seen be the intruder.

- *Temporal presence*
  In common with the omnipresence often assumed, one also often assumes that the intruder has always been there and that it will always continue to be there. This is compatible with the idea that the intruder is the network.
- *Physical intrusion*
  A common assumption is to have an adversary that does not physically compromise any principal entity. This assumption is useful in analysis, but is often wrong in the real world.
- *Dishonest principal*
  Most intruder models, but not all, consider the principal parties to be honest. Of course, this cannot be relied upon in the real world. The technical assumption is often full mutual trust, but trust may be betrayed. We also have that while the principal intends to be faithful, he/she can be fooled and manipulated. Thus, he/she may not be trustworthy.
- *Social engineering*
  Phishing scams may be considered a social engineering attack. A surprising number of people fall victim to such attacks, but they are rarely accounted for in intruder models.

3.3 Differences between MIX Networks and Onion Routing

We now briefly highlight the main privacy services offered by MIX networks and Onion Routing systems. The Tor project [79] and Onion Routing, in general, are to a large degree due to Paul Syverson, and we let him spell out the principal differences between MIX networks and Onion Routing:

> Mix networks get their security from the mixing done by their component mixes, and may or may not use route unpredictability to enhance security. Onion routing networks primarily get their security from choosing routes that are difficult for the adversary to observe, which for designs deployed to date has meant choosing unpredictable routes through a network. And onion routers typically employ no mixing at all. This gets at the essence of the two even if it is a bit too quick on both sides. Mixes are also usually intended to resist an adversary that can observe all traffic everywhere and, in some threat models, to actively change traffic. Onion routing assumes that an adversary who observes both ends of a communication path will completely break the anonymity of its traffic. Thus, onion routing networks are designed to resist a local adversary, one that can only see a subset of the network and the traffic on it.

- Paul Syverson in "Why I'm not an Entropist" [233]

There are different implementations of MIX networks and OR systems, and these offer somewhat different services. We have tried to avoid too much specific discussions here. This particularly applies to our treatment of MIX network. With respect to Onion Routing, we basically discuss the second generation Tor approach. So what we have then is:

MIX – Messages are "mixed" by chained MIX components.

The MIX components main purpose is to permute the input/output such that it is hard to guess which outputs correspond to a given input. The MIX chain will then, much like rounds in a cipher functions, makes analysis less and less tractable. The adversary is assumed to be able to listen in to all traffic and may also modify traffic (in some models). To avoid traffic analysis, the MIX components will aggregate messages and may introduce non-deterministic delays. Full anonymity is, in theory, possible. The communications model is packet switched.

OR – Messages are forwarded along predetermined, but secret routes.

The adversary will not have complete control and can in particular not watch all end-points. The primary mechanisms is to conceal the route from the adversary. Full anonymity is generally a non-goal. The communications model is circuit switched, and a circuit must be established prior to sending user plane traffic. Messages are not delayed and OR is well suited for real-time and interactive traffic types.

Our privacy cases match MIX networks and OR systems roughly. Case A can be addressed by a mixnet. Alice will be able to communicate anonymously and Bob will not know who Alice is. The communications may have high latency. Case B can be addressed by an OR system. Alice and Bob will be able to communicate anonymously and interactively.

3.4 Brief Introduction to MIXes

3.4.1 The Idea of MIX Networks

MIXes, or rather MIX networks, were invented by David Chaum in 1981 [57]. A Mix network (mixnet) is a multistage system. For each stage one takes a batch of messages (packets) as input. The messages are then cryptographically transformed and the sequence permuted, before they are forwarded towards the next stage The goal is to achieve untraceability between the input and output batches. The purpose of the mixnet is to create an apparently random and untraceable communications path by using a chain of MIX servers. An important point here is that each message is forwarded through the MIX networks individually. So if Alice sends **N** messages belonging to a message stream through the mixnet, then there should be no observable pattern to the messages as they proceed through the mixnet. That is, the messages would likely follow very different routes and they would likely arrive out-of-order at Bob. "Randomized" routing is not by itself sufficient. One must also conceal the sender and receiver addresses such that an intruder cannot easily reconstruct the routes by collecting address information (in/out) of the individual MIX units. Therefore, each message is encrypted by each MIX unit using public key cryptography. The encryption is layered like a Russian Matryoshka doll with the senders original message as the innermost layer. Each MIX server decrypts and removes its own layer of encryption to reveal where to send the message next.

This scheme with multiple encryptions initially and then successive route-and-decrypt steps is not the only possible approach. One may also construct MIX networks that follow a recipe with re-encryption at the MIXes. That is, MIX_N decrypts the message, decide on the next MIX (MIX_{N+1}), then re-encrypts the message before forwarding the message to MIX_{N+1}. Note that a lot of care needs to be taken with respect to traffic analysis if the untraceability is to have any credibility.

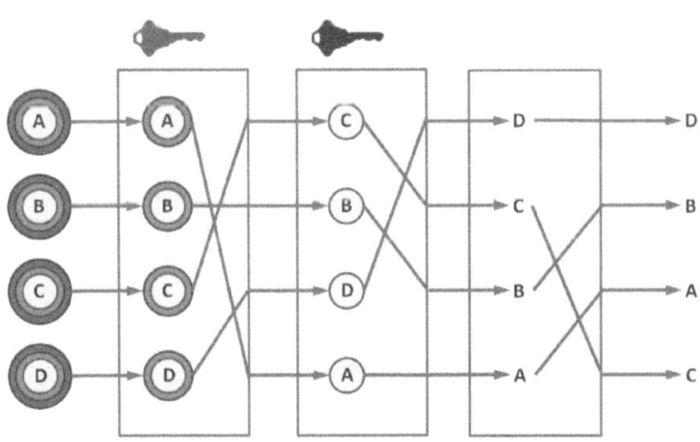

Figure 3.4 Cascade MIX Network with 3 MIXes (source: Wikimedia Commons)

Figure 3.4 depicts a simple decryption MIX network. The messages are encrypted with public keys belonging to the MIXes. When a MIX receives a message, it decrypts the message with its private key and forwards the message toward the next MIX. The MIXes may delay messages and rearrange the order of messages it receives before transmitting the messages. Mixnets may provide two-way communications and this requires that the receiver (Bob) is able to reply to the sender (Alice).

3.4.2 Mixnet Uses

It may be debatable if electronic voting is a good idea, at least for general elections since there are many important threats and substantial risk. However, one of the possible applications of mixnets is in secure and anonymous electronic voting. In e-voting protocols, the mixnet anonymously communicates the ballots from the voters to the electoral authority. With a deep enough mixnet (number of stages) it should be possible to provide the necessary ballot secrecy with an adequate degree of robustness, public trust and confidence [150]. Elections, electronically conducted or not, should ideally be error-free and problems related to voter fraud and coercion have to be resolved. Credible counting verifiability must be provided. More on electronic voting can be found in [61, 182]. The book “Towards Trustworthy Elections” [58] is a must read for anyone seriously interested in electronic voting.

Other applications such as anonymous e-mail can also benefit from using mixnets. It is not apparent that anonymous email is always a good idea (consider spam), but technically mixnets will do the job. There are other applications that require anonymity too, but we note that generally speaking one cannot easily have real-time communications over mixnets. That is, there are mixnets that do support low-latency communications, but there are then limitations to the anonymity they can provide since timing information will inevitably leak. This may allow the intruder to reconstruct the path. This is true for the passive intruder and even more so for the active intruder.

3.4.3 Mixnet Topologies

3.4.3.1 Synchronous Cascade Mixing

Figure 3.4 depicts a simple cascade topology. All messages will pass through each fixed stage in order. The messages may arrive during an interval at the first stage, but then generally the subsequent batch forwarding will be more-or-less synchronous. To achieve credible mixing, one needs a fair amount of message. With **m** input messages even a blind guess on the output will be right

1/m of the time (per stage). Clearly, more inputs provide better anonymity. Likewise, more stages provide better anonymity.

3.4.3.2 Free-route Mixing

So-called Free-route mixing comes in two varieties, the synchronous one and the asynchronous one. Free routes basically mean that the different messages in a message stream will take different routes through the mix network as depicted in Figure 3.5. Complex topologies obviously make fully synchronous communications difficult and so free-route mixnets are typically asynchronous. The free-route paths must be directed and there should be no cycles. Unbalanced link and different link latencies are parts of this picture. We recommend the article "A Survey on Mix Networks and Their Secure Applications" [222] for a broad discussion on different mixnets and there advantages/disadvantages.

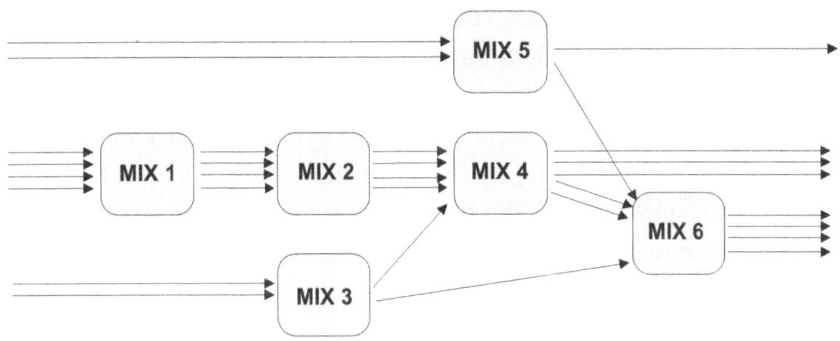

Figure 3.5 Free-route MIX Network

3.4.4 Mixing Strategies

We may classify the MIX strategies not only according to the network topology used but also on how they perform the cryptographic concealment. Basically, we have these options:

- Decryption mixnets
- Hybrid mixnets
- Re-encryption mixnets

3.4.4.1 Decryption Mixnets

This category of mixnets is the original one outlined in [57]. In decryption mixnets, the sender must encrypt the message with the keys for all the stages.

This is shown in Figure 3.4. The cryptographic transform will literally transform the message such that there is no apparent relationship between message **M** in the input and message **M'** in the output, although the payload of **M** and **M'** is the same.

When we say that the payload is the same we are cheating a little. The "payload" is relative and the decryption-and forwarding process will include mixnet protocol data etc. This also means that unless padding is used the message will shrink as it passes the different stages. The analogy "to peel an onion" is quite illustrative of the process. Traditionally, one has used public-key cryptography, but it is also possible to devise symmetric key solutions. There is a lot of technicalities about how to construct the MIX onions and how the protocol should work. For a detailed account, we refer to [222] for an overview of MIX networks and their properties.

3.4.4.2 Hybrid Mixnets

The typical decryption mixnet relies heavily on public-key cryptography. This is nice in theory, but unfortunately quite inefficient in reality. Thus, there is a need for mixnets that are less resource wasteful. There are several different types of hybrid mixnets, but they all use symmetric key cryptography in place for public-key cryptography. Some of the solutions effectively set up a route and gather symmetric keys by means of public-key operations. These solutions will resemble Onion Routing to some degree.

3.4.4.3 Re-encryption Mixnets

The decryption mixnets have some general properties that could be problematic.

1. The onion size will decrease along the path (per stage).
2. The onions will, in principle at least, be traceable by size.
3. The sender must encrypt for all stages before sending the message.
4. The path must be predetermined (per onion).

There are ways around this in decryption mixnets, using ElGamal-based encryption, but these are fairly complex solutions [208] and they have their own share of weaknesses. Re-encryption mixnets aim to address these aspects simply by allowing the MIX stages to re-encrypt the onion. This can be done by using ElGamal encryption properties [88]. Basically, we have:

$$E_K(m,r) = (g^r g^{r'} || m K^r K^{r'}) = (g^{r+r'} || m K^{r+r'}) = E_K(m, r + r')$$

Where m is the message, K is the public key, r, r' are random strings and g is the generator. By re-encrypting as shown with a random string r', the appearance of the encryption of m is completely changed. Decryption is by means of the private key and the sum of the different r-values instead of the initial r. This "additive" property is also the basis for homomorphic properties of the ElGamal crypto-system. See Chapter 4 for more on homomorphic cryptography in SMC.

3.5 Brief Introduction to Onion Routing

3.5.1 Onion Routing

Onion routing is a method for achieving anonymous communications over the open internet. The service is somewhat analogous to mail delivery by the Post Office, in which the sender might be anonymous with respect to the receiver and with respect to any external parties. In the Post Office analogy, one still needs to trust the Post Office and one cannot attain "traffic analysis" privacy with respect to an adversary that monitors all postboxes (global observer). The use of envelopes provides data privacy with respect to mail contents, but privacy against traffic analysis may often be equally important and may in many circumstances be a prerequisite for sending a message in the first place.

An onion router network is an overlay network consisting of a set of onion routers (OR). Onion router networks, such as the Tor network, provide connection-oriented cryptographically protected services through the set of onion routers. The services aim at providing low latency connectivity and are organized around circuits and streams of data traversing the circuits. The Tor circuits are "virtual" and each onion router only knows the two nearest neighbors in the circuit.

The name "onion routing" gives associations towards an onion with different layers of information that might get peeled off during transmission. Another analogy is the "telescoping" in which one builds the path incrementally and interactively during circuit setup. The Tor network is the current state-of-the-art with respect to an actually deployed onion network. In "A Peel of Onion" [232] Paul Syverson, one of the inventors of Onion Routing, provides a personal overview of OR/Tor, its intended applications and services. The freely available paper "Tor: The Second-Generation Onion Router" [79] provides a good introduction to Tor.

3.5.2 Onion Routing in Practice - The Tor Approach

Tor is an abbreviation for "The Onion Router" network and Tor is indeed a network of onion routers and of virtual circuits that allows people and groups to improve their privacy and security on the internet. Onion routing relies on multiple layers of security that are peeled (like onion skin) one by one as a message is routed through the Tor network. The primary resource for Tor is the Tor homepage at `https://www.torproject.org/` and much of material in this clause is directly inspired by documentation found on the Tor homepage. You are strongly encouraged to visit the Tor homepage for more information and further details.

3.5.2.1 Onion Routing

The Tor network is an overlay connection-oriented mesh network where each onion router maintains connectivity to every other onion router. The connections between ORs are secured by means of the Transport Layer Security (TLS) security protocol [78]. The OR users run a so-called onion proxy (OP) to fetch directories, establish circuits across the OR network, and handle connections from user applications. The onion router on the other side of the circuit handles the connection when it exits the OR system by forwarding the data (IP packets) towards the requested destinations and relays data back again. Figure 3.6 depicts a simple OR network, but here the secured interconnect tunnels are not shown. Each user locally runs a so-called Onion Proxy (OP). Some of the ORs will act as directory servers, providing signed directories of known routers in the network. The directory servers should be explicitly trusted. Client A (Alice) will must start any session building a virtual circuit, and for this she will need access to the directory services to obtain data about which ORs are available, exit policies etc.

3.5.2.2 Circuits and Cells

The users who wish to use the OR network first choose a path through the network and build a circuit. The setup phase consist of "telescoping" the key agreement part through the OR network. This part of the protocol uses TLS to set up each leg of the circuit. This may be considered the *control plane* signalling of the protocol. In each circuit, the respective OR knows the path to its predecessor and successor, but importantly an OR does not know any other nodes in the circuit. Having set up the circuit the client may start to use the path. This will be the *user plane* part of the protocol. Data traffic passes through the circuits in units called cells.

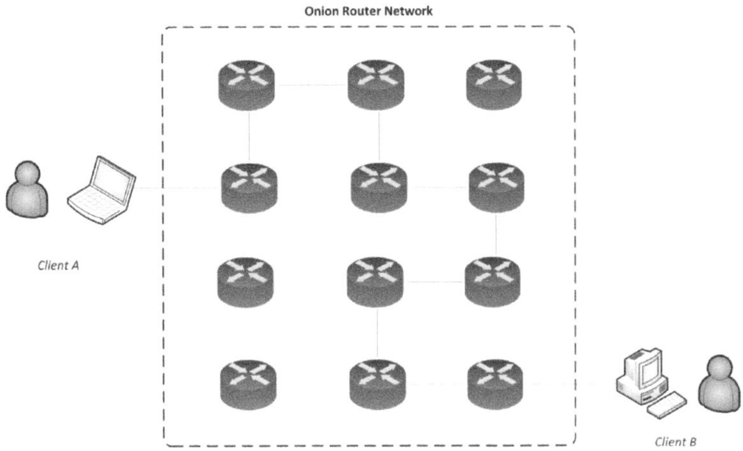

Figure 3.6 Basic Onion Routing

The cell has a fixed-size and the cells are encrypted during transit between the ORs. The encryption key is changed for each leg in the circuit – somewhat analogous to changing envelopes between intermediate post offices. The encryption uses throw-away keys (ephemeral keys) with perfect forward secrecy to avoid compromises to propagate. Each cell is exactly 512 bytes wide, and consists of a header and a body (payload). The header includes a 2 byte circuit identifier (`CircID`) which specifies the circuit the cell belongs to. Figure 3.7 depicts a simple OR network with two circuits. The cell header also contains a command byte (`CMD`), which specifies how to interpret the payload and thereby what to do with the payload.

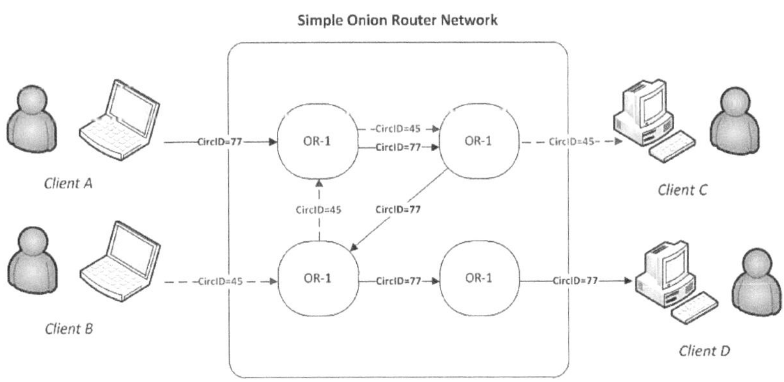

Figure 3.7 Simple Circuits

3.5.2.3 Control Plane and User Plane Cells

The cells are either control cells (control plane) or they are relay cells (user plane). The control plane messages are *padding*, *create/created* and *destroy*. Relay cells include a relay header indicator (`Relay`), a stream identifier

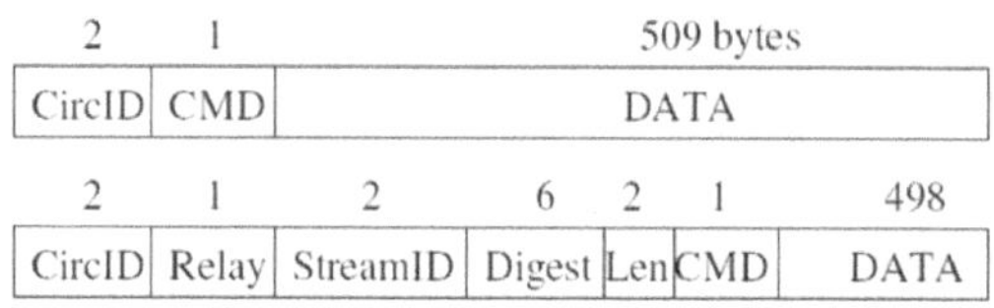

Figure 3.8 Cell layout

(`StreamID`), an end-to-end integrity checksum (`Digest`), a payload length indicator (`Len`) and a relay command (`CMD`). The `Len` indicator is respective to the amount of data within the `DATA` field. Many streams may be multiplexed over one circuit. The relay cell commands are targeted at controlling data streams.

3.5.3 A Tor Example

The following example is according to the second generation Tor networks. We have the following components to such a systems:

- The Tor client (Alice)
- Directory Service
- Onion Routers (OR1, OR2, OR3, ..., ORn)
- Target site (Bob)

Figure 3.9 depicts a Tor setup in a two-hop configuration. The figure is from [79] and shows circuit setup to allow Alice to privately browse a web page. For the sake of simplicity, the example only contains two ORs. In a real-world scenario, one would want a longer path to ensure that the adversary cannot predict the path. The control plane circuit setup is given below:

1. `Directory lookup`
   The client proxy interrogates the directory service to a list of Tor nodes.
2. `Circuit setup: client to OR1`
   This establishes the first circuit $c1$. Symmetric keys are now shared between the client and OR1 (K_{or1}).
3. `Circuit setup: client to OR2 (tunneled via OR1)`

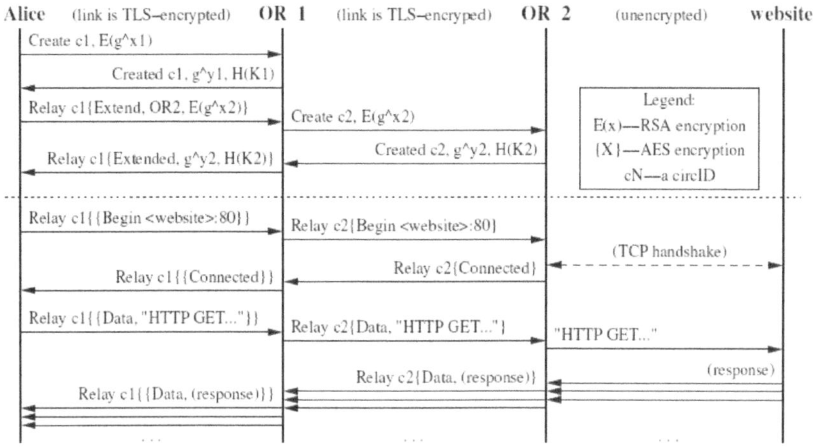

Figure 3.9 Tor setup of a two-hop circuit

This establishes the second circuit $c2$. Symmetric keys are now shared between the client and OR2 (K_{or2}).

After the circuit has been established the browsing can commence. The "`Begin <website>:80`" from Alice is forwarded into the two-layered (encrypted) tunnel. The number of layers reflects the number of ORs involved in setting up the path. The outer layers are successively removed along the path in a "telescoping" way. As is seen in figure 3.9 the connection from the last OR towards the target site is unencrypted.

3.5.4 Low-latency May Mean Trouble

Low-latency services are part of the Tor design. They are problematic since one may reconstruct the path by closely monitoring the inter-OR communications. Some of these aspects of Tor are discussed in "Deploying low-latency anonymity: Design challenges and social factors" [80] by Dingledine, Mathewson and Syverson (the principal designers of Tor) and in the paper "How much anonymity does network latency leak?" [133]. The conclusions are quite clear that Tor and related approaches are fairly vulnerable to timing attacks. The attacks are not trivial to execute, but they are effective. Active attacks are the worst and in [108] one attempts to address this problem by means of a "black-box" padding scheme. The cost is that there may be more latency and there will be an increase in bandwidth requirements.

3.5.5 Other Attacks

There are many possible attacks on Tor systems and one of them is based on triggering congestions, causing loss of availability (Denial-of-Service). These attacks require long paths and are to some degree quite effective. The purpose of the long paths is to consume resources in the network. In [103] this is demonstrated by using circuits that loop onto themselves. One may address this partially by requiring paths to be directed and acyclic, but it is generally hard to prevent this type of attack if long paths are allowed.

3.5.6 Analysis

Several papers have been written analyzing of Tor's performance, security and anonymity properties. The paper "A formal treatment of onion routing" [51] contains a formal analysis of onion routing. The paper starts off by defining a model of onion routing and uses this for deep mathematical analysis of the security/privacy properties. In the paper "Trust-based anonymous communication: Adversary models and routing algorithms" [153], the authors analyze the trust aspects of anonymous communications and looks into the intruder models. The work is interesting and practical since one may very well assume that the intruder sets up its own ORs in the network. In particular, when the intruder controls many ORs it turns out that the routing algorithms play a significant role in the actually achieved anonymity. The trust analysis is also applicable to MIX networks since the intruder can just as well take control over a MIX stage. Data mining as a method may be used for analysis of OR and MIX networks too. Correlation of user behaviors may leave a pattern in the observed (encrypted) traffic. In [109], the authors provide this type of analysis of OR in a black-box model.

3.6 Summary

Both MIX networks and OR systems can provide anonymous communications, but neither approach is without problems. From a practical perspective, Tor is far ahead of anything else, providing end-user software for many different platforms and having a large network of onion routers. While Tor may appear to provide lesser services than MIX networks, the pragmatic approach has also meant that Tor, with its weaknesses, is capable of providing real privacy to people. It may not withstand an all-powerful intruder, but it does provide reasonable privacy for whistleblowers, activists, businesses and even for military and law enforcement agencies. Needless to say, but it may of course be abused, and this is unfortunately the price to pay for providing credible privacy to people. The ideas behind MIX network and OR systems are interesting and important and this, together with explanations of anonymity properties, have been our focus in this chapter. We haven't gone too deep into the technical details of MIX and OR in this chapter, but suffice to say that the devil hides in the details and even the most "provably secure" design can be betrayed by omissions or errors here. No amount of advanced mathematics can help out when the premises are wrong and to label something "provably secure" has often limited practical value.

4

Secure Multi-Party Computations and Privacy

It is widely acknowledged that technology alone is not a panacea for protecting privacy and legal protection is also needed. However, it is also clear that without technological enforcement it is difficult to provide reasonable privacy protection. Therefore, various technological solutions providing different degree of privacy protection have been proposed recently. In the following chapter, we will discuss some of such solutions and underlying technologies.

This chapter informally introduces the idea of secure multi-party computations (SMCs) and demonstrates their applicability to privacy protection. We outline SMCs and provide some examples to show how SMCs can be used to provide specific privacy preserving solutions of some known problems with special focus in the area of telecommunication systems.

4.1 Motivation

People use telecommunication-based applications daily and the system collects a large amount of information related to their activities. Such telecom networks create opportunities for performing cooperative tasks based on computations with inputs supplied by separate users. Moreover, such computations could be performed even among mutually distrusted parties. Since in many cases users' input contain private information reflecting users daily activities (e.g., travel routes, buying habits, etc.), secure multi-party computations become a relevant approach for dealing with privacy in such applications. Many applications have to utilize available private data to improve the quality and security of every day life. For example, private data have been used to develop such important applications as traffic jam monitoring, monitoring of elderly people, anti-terror related monitoring of suspects, etc. Meanwhile, with an increased amount of private data collected in telecommunication networks, privacy concerns become a critical issue. Some questions that should be addressed include how exactly the private data will be used;

can the data be misused to invade people's privacy; and who is watching the watchers.

Many techniques have been proposed to deal with these privacy issues such as, for example, k-anonymity [231], data transformation/randomization, etc. In this chapter, we focus on techniques that are based on secure multiparty computations.

4.2 Secure Multi-party Computations: An Informal Introduction

Secure multiparty computation (SMC) is a cryptographic technique that enables computations on data received from different parties in such a way that each party knows only its own input and the result of the computations.There are many SMC-based solutions that can be used to ensure privacy preservation and protection. Informally, they can be described as a computational process where two or more parties compute a function based on private inputs. Privacy in this context means that none of the parties wants to disclose its own input to any other party.

Formally, a secure multi-party computation problem can be formulated as follows. Assume that there are n parties ($n > 1$) that want to perform some computations jointly. It means that each party is willing to contribute some data to these computations. However, each party is willing to contribute its input only privately, that is, without disclosing it to the other participating parties or to any third party. Generally, this problem can be seen as a computation of a function $f(x_1, x_2, ..., x_n)$ on private inputs $x_1, x_2, ..., x_n$ in a distributed network with n participants where each participant i knows only its own input x_i and no more information except output $f(x_1, x_2, ..., x_n)$ is revealed to any participant in the computation [123].

Historically, the secure multi-party computation problem was first introduced by Yao [253] where a solution of the so-called Yao's Millionaire problem was proposed. (The problem that was formulated is about how two millionaires can learn who is richer without revealing any information about their net worth.)

According to theoretical studies, the general SMC problem is solvable based on the circuit evaluation protocol [122]. At the same time, it was observed that such solutions are not practical from the efficiency point of view. Therefore, finding efficient problem-specific solutions was recognized as an important direction for future research and many specific solutions were proposed in the literature recent years. The authors have proposed more efficient solutions for problems in the areas of data mining, information retrieval, computational geometry, statistical analysis, etc [43, 84, 85, 212, 241].

Initially, research on such problem-specific solutions was performed under the assumption of the ideal security model that assumes zero information disclosure. However, despite the fact that the specific solutions for problems mentioned above are more efficient than general solutions, they are still resource demanding and not always very usable in practical applications with constrained resources (e.g., energy, computational power, or communication broadband). At the same time, many authors started to notice that for some applications, users would be willing to accept a reduced level of security if they could achieve sufficient efficiency, especially in the cases where ideal security solutions are not usable because of unacceptable performance. It has been shown that in this case, the main goal is to achieve efficiency with a sufficient level of security, but not security itself.

This is the main reason why recent research is focused on a new, more practical paradigm utilizing a security model permitting some information disclosure. By relaxing the security imposed restrictions, more practical solutions with better performance may be designed. In the following sections, we briefly describe some of such solutions.

A simple example of an efficient SMC that illustrates the idea of privacy preserving computations is the secure sum protocol. We assume that there are n parties $P_0, P_1, ..., P_{n-1}$ such that each P_i has a private data item d_i, $i = 0, 1, ..., n-1$. The parties want to compute $\sum_{i=0}^{n-1} d_i$ privately, that is, without revealing their private data d_i to each other. The following solution was presented in [68] and solves the problem described above. We assume that $\sum_{i=0}^{n-1} d_i$ is in the range $[0, m-1]$ and P_t is the protocol initiator. At the beginning P_t, chooses a uniform random number r within $[0, m-1]$. Then P_t sends the sum $d_t + r \bmod m$ to the party $P_{t+1 \bmod n}$. Each remaining party P_i does the following: upon receiving a value x the party P_i sends the sum $d_i + x \bmod m$ to the party $P_{i+1 \bmod n}$. Finally, when party P_t receives a value from the party $P_{t-1 \bmod n}$, it will be equal to the total sum $r + \sum_{i=0}^{n-1} d_i$. Since r is only known to P_t, it can find the sum $\sum_{i=0}^{n-1} d_i$ and distribute it to other parties.

4.3 Basic Techniques and Building Blocks

There are some basic building blocks that are repeatedly used in many solutions proposed in the literature. Before we present privacy preserving solutions we will briefly review some of these building blocks in this section. Almost all solutions considered later in this chapter are based on using these basic building blocks. We provide only a short overview of some of the most popular techniques and give recommendations for further reading.

4.3.1 Yao's Millionaire Problem

This problem is important in the context of data mining and e-commerce, e.g., in such applications as on-line bidding and auctions. It was considered by Yao in [253]. It contains the first scheme for secure comparison. The scheme assumes that there are two parties, Alice and Bob, where Alice has a number a and Bob has a number b describing their wealth. Alice and Bob want to verify whether $a \leq b$ or $a > b$ without revealing information about a and b to each other. That is, two millionaires, Alice and Bob, wish to know who is richer without revealing their actual wealth.

The proposed solution assumes that a public key crypto system is available where E_A denotes encryption with Alice public key and D_A denotes decryption with Alice private key. Actual wealth of Alice and Bob are presented by $a, b \in \{1, 2, ..., n\}$. The proposed solution (known as Yao's protocol) describes privacy preserving calculation of f:

$$f(a,b) = \begin{cases} 1, & if\ a \leq b \\ 0, & if\ a > b \end{cases}$$

Yao's protocol can be presented as follows:

1. Bob picks a random N-bits integer x, and computes $k = E_A(x)$;
2. Bob sends Alice the number $(k - b + 1)$;
3. Alice computes privately $y_u = D_A(k - b + u)$ for $u = 1, 2, ..., n$;
4. Alice generates a random prime p of $N/2$-bits, and computes $z_u = y_u \bmod p$ for all u; if all z_u are differ by at least 2, stop; otherwise generate new p and repeat process until all z_u differs by at least 2;
5. Alice sends $z_1, z_2, ..., z_a, z_{a+1} + 1, ..., z_n + 1$ (all $\bmod p$) and p;
6. Bob looks at the b number sent by Alice, and decides that $a \geq b$ if it is equal to $x \bmod p$, and $a < b$ otherwise.
7. Bob tells Alice the result.

The details analysis and explanation can be found in [253].

4.3.2 Homomorphic Encryption

Homomorphic encryption is a form of encryption that permits execution of a specific algebraic operation (denoted here as $\oplus$) on the plaintext by executing a (possibly different) algebraic operation (denoted here as $\bullet$) on the corresponding ciphertext. The homomorphic cryptosystems are an important basic building block in many secure multiparty protocols. Several such cryptosystems have been proposed in the literature [37, 88, 191, 206].

More formally it can be defined as follows. Let us consider a public-key cryptosystem with the homomorphic property where encryption and decryption are denoted as $E\,()$ and $D\,()$ respectively. The additive homomorphic cryptosystem provides operation $\bullet$ on encrypted data corresponding to addition operation on the cleartext (that is, $\oplus$ denotes addition): $E(m_1) \bullet E(m_2) = E(m_1 + m_2)$. The multiplicative homomorphic cryptosystem provides operation $\bullet$ on encrypted data corresponding to multiplication operation on the cleartext (that is, $\oplus$ denotes multiplication): $E(m_1) \bullet E(m_2) = E(m_1 \times m_2)$.

Thus, by using additive homomorphic cryptosystem, one can find encrypted sum of encrypted x and y, i.e., $E\,(x) \bullet E\,(y) = E\,(x + y)$. Consequently, since $E\,(yx) = E(\underbrace{x + x + \cdots + x}_{y}) = \underbrace{E\,(x) \bullet E\,(x) \bullet \cdots \bullet E\,(x)}_{y}$, we are able to multiply encrypted data if only one of the multipliers is encrypted.

As a simple example of a homomorphic cryptosystem, we can consider the RSA cryptosystem. It is easy to see that $E\,(x_1) \bullet E\,(x_2) = (x_1^e \bmod n)\,(x_2^e \bmod n) = x_1^e x_2^e \bmod n = (x_1 x_2)^e \bmod n = E\,(x_1 x_2)$, where (e, n) is a public key. In this case, both $\oplus$ and $\bullet$ are modular multiplication.

RSA is an example of multiplicative homomorphic cryptosystem, that is, given only the public-key and the encryption of m_1 and m_2, one can compute the encryption of $m_1 \cdot m_2$. Assume that the RSA public key is (e, n) where n is the modulus and e is the exponent and encryption of a message m is given as $E(m) = m^e \bmod n$. The homomorphic property is following: $E(m_1) \cdot E(m_2) = m_1^e m_2^e \bmod n = (m_1 m_2)^e \bmod n = E(m_1 \cdot m_2)$

The Paillier cryptosystem [206] is an example of an additive homomorphic cryptosystem, that is, given only the public-key and the encryption of m_1 and m_2, one can compute the encryption of $m_1 + m_2$. Assume that (n, g) is a public key where n is the modulus and g is the base, and $E(m) = g^m r^n \bmod n^2$ is the encryption of a message m for some random

$r \in \{0, 1, ..., n-1\}$. The homomorphic property is following: $E(m_1) \cdot E(m_2) = (g^{m_1} r_1^n)(g^{m_2} r_2^n) = g^{m_1+m_2}(r_1 r_2)^n = E(m_1 + m_2 \bmod n)$.

For many applications it would be essential to have cryptosystems that support both addition and multiplication of encrypted data. Such cryptosystems are known as fully homomorphic cryptosystems. Only recently, such a cryptosystem was proposed [119]. However, to provide efficient support for both operations is a difficult problem that has not solved yet.

4.3.3 Oblivious Transfer

Oblivious transfer describes communication between two parties, sender and receiver, where the sender transmits part of the data to the receiver. The receiver chooses a part of the received data in a privacy protecting manner: the sender is assured that the receiver gets no more information to which it is entitled. The sender learns nothing about which part of the data has been selected. The 1-out-of-N oblivious transfer is a method that allows a party to access one of the N secrets, without getting any information about remaining secrets and without disclosing which of N secrets was accessed. Formally, 1-out-of-N oblivious transfer protocol provides a method where Bob has N items $x_1, ..., x_N$ and Alice learns one of the items, x_i of her choice, without learning anything about the other items, and without allowing Bob to learn anything about i.

The first such protocol was proposed by Rabin [214] and since then several variants of different types of oblivious transfer protocols were proposed (for example, [104, 192]). They served as important building blocks for many cryptographic applications such as protocols for signing contracts, certified email or flipping a coin over phone [104]. Theoretically, it was shown by Kilian [165] that by using only oblivious transfer, it is possible to construct any secure protocol.

Here, we present an example of efficient 1-out-of-2 oblivious transfer (proposed by Even, Goldreich and Lempel in [104]):

1. Alice has messages m_0 and m_1 and wants to send one of them to Bob without knowing which Bob receives;
2. Alice generates RSA key pair (e, N) and (d, N);
3. Alice generates random values x_0 and x_1 and sends to Bob x_0, x_1 and (e, N);
4. Bob picks up x_b;
5. Bob generates a random k, and sends to Alice $v = (x_b + k^e) \bmod N$;

6. Alice calculates $k_i = (v - x_i)^d \bmod N, i = 0, 1$, one of these k_i is equal to k;
7. Alice blinds the two secret messages with possible keys k_0, k_1, $m_i' = m_i + k_0, i = 0, 1$ and sends m_0',m_1' to Bob;
8. Bob knows which of two messages he can unblind with k, so he computes $m_b = m_b' - k$

However, oblivious transfer computational requirements are quite demanding in terms of resources and they often became the bottleneck in many applications that use them. Therefore, finding efficient solutions is an important research area and many such solutions have already been proposed [192].

The problem called Private Matching presented in the next subsection can be also seen as an example of 1-out-of- N oblivious transfer protocol.

4.3.4 Private Matching of Interests

The objective of the following simple example is to illustrate how privacy can be achieved. More advanced examples and discussions can be found in [113].

We assume that Alice has a private set $A = \{a_1, a_2, ..., a_n\}$ and Bob has a private set $B = \{b_1, b_2, ..., b_m\}$. The protocol that follows is a two-party protocol between Alice and Bob that finds the private intersection on their inputs. Alice defines a polynomial $P(x)$ whose roots are her private set A as follows:

$$P(x) = (a_1 - x)(a_2 - x) \cdots (a_n - x) = \sum_{i=0}^{n} \alpha_i x^i.$$

Let E represent an additive homomorphic encryption. Alice sends $E(\alpha_0), E(\alpha_1), \ldots, E(\alpha_n)$ where $E(\alpha_i)$ is a homomorphic encryption of α_i. Bob evaluates polynomial $P(b_i)$ by finding $E(P(b_k)) = \sum_{i=0}^{n} E(\alpha_i) \cdot b_k^i$. Since E represents an additive homomorphic encryption, Bob can calculate $E(P(b_i))$ without knowing the real values of coefficients α_i. After that, Bob selects a random number r and calculates $E(r \cdot P(b_k) + b_k)$. If b_k is in A, then $P(b_k) = 0$ and $E(r \cdot P(b_k) + b_k) = E(b_k)$. Therefore, Alice can find whether b_k is in intersection $A \cap B$ by decrypting $E(b_k)$ with her private key and checking that $D(E(b_k)) = b_k$ is in A. If b_k is not in A then $P(b_k) \neq 0$ and the result of $D(E(r \cdot P(b_k) + b_k))$ is random. Thus, Alice learns whether b_k is in A without revealing A to Bob and without learning b_k when b_k is not in A.

4.4 Examples of SMC-based Privacy Preserving Solutions

The main purpose of this section is to provide some illustrative examples of broad applicability of secure multi-party computations to privacy preservation to stimulate design of new privacy preserving solutions.

4.4.1 Private Information Retrieval

The problem of Private Information Retrieval (PIR) can be formulated as follows. Assume that a user wants to query a database in a private way, that is such that the database receives no information about the query. Formally, we consider the database as a n-bit string, $x = x_1x_2 \cdots x_n$ and the query is the bit i. PIR means that the user can retrieve the bit x_i by sending the query i such that database learns no information about i. The problem was introduced in [63] and since then studied intensively in literature (see, for example, [36, 62]). Since the straightforward solution to PIR would be to send the whole database to the user, the main goal of this research was to minimize sub-linear communication complexity. However, PIR, as it was introduced in [63], is not concerned with privacy of the database. The extension of the problem introduced in [120] and called symmetrically private information retrieval (SPIR) protects privacy of the database (in addition to user privacy) where database privacy means that the user cannot obtain more information about the database except contained in the result of her query. SPIR can also be seen as very similar to oblivious transfer discussed in the previous sections.

4.4.2 Selective Private Function Evaluation

This problem was introduced in [55] where several solutions were presented. The problem is formulated as follows. We assume that several servers hold copies of a database $x = x_1x_2 \cdots x_n$. A client chooses a function f and indices $i_1, i_2, ..., i_k$, and interacts with servers (one or more) in order to compute $f(x_{i_1}, x_{i_2}, ..., x_{i_k})$ privately in the sense that servers know nothing about chosen indices.

An example of problem setting that illustrates usability of selective function evaluation can be described in the following way. Consider a scenario where there is a database containing both public information about users (e.g., their addresses and phone numbers) and private information (e.g., age, salary and mobile phone using habits of each user). Public information is freely accessible, but private information is sensitive and should not be accessible. For example, a company wants to perform some statistical analysis on a selected subset of private data without revealing selection criteria while the database owner wants to reveal only data which will be used for analysis.

4.4.3 Privacy Preserving Scientific Computations

Many industries, including the telecommunication industry, have to solve problems related to planning, routing, scheduling or optimization. These problems are often modeled as systems of linear equations or linear least squares problems. However, in the scenario in which two or more untrusted parties (for example, competitors) want to solve the problems without revealing private data, traditional well-studied approaches are not applicable.

Consider for example a scenario in which two telecommunication companies want to optimize joint use of their networks without revealing proprietary information about internal structure, constraints, etc. In many cases, such problems involve solving systems of liner equations where none of the parties has knowledge about all equations. The privacy preserving case of the problems that have been considered in the literature [83] can be formalized in the following way. Assume that Alice has m private linear equations and Bob has $n - m$ private linear equations represented by $M_A x = c_A$ and $M_B x = c_B$ respectively, where x is an n-dimensional vector. The authors show how Alice and Bob can jointly find a vector x that satisfies all equations without revealing to each other any information about their own equations.

4.4.4 Computational Geometry Problems

Several computational geometry problems in the privacy preserving setting have been considered in the literature [33]. They include point inclusion in polygons, polygon intersections, finding the closest pair of points, etc. Many of these problems can be easily interpreted in the context of telecom applications [179].

The simplest of the computational geometry problems is the so-called privacy preserving Point-Inclusion Problem [33]. The problem can be formulated as follows: assuming that Alice has a point x and Bob has a polygon P, determine whether x is inside P without revealing to each other any information about the relative position of x with respect to P.

A privacy preserving polygon intersection problem can be formulated as follows. Assume that Alice has a polygon P_A and Bob has a polygon P_B. Both Alice and Bob want to find out whether $P_A \cap P_B$ is empty without revealing any information about the polygons to each other.

Finally, assume that Alice has a set S_A of points in the plane and Bob has a set S_B of points in the plane. The privacy preserving closest pair problem is about finding (by both Alice and Bob) pairs of closest points among points in $S_A \cup S_B$ without revealing to each other any information about S_A and S_B.

4.5 Some Potential Real-Life Applications

Secure multi-party computations have very wide variety of potential applications. As we already illustrated, many classical computational problems can be reformulated in a privacy preserving manner. In this section, we give examples of some real-life applications that have been considered in the literature.

4.5.1 Privacy Preserving Location

Preserving the privacy of a user is an important challenge for mobile and wireless applications. The purpose is to utilize user location without actually disclosing it either to a service provider or to any third party. One possible approach would be to use the privacy preserving solutions for computational geometry problems [33]. In the context of telecommunication systems, the location and identity privacy of the current 2G/3G systems have been analyzed in [177–179]. Authors argue that using Identity-Based Encryption is ideal for fast set-up in new roaming areas. They proposed solutions for spatial control and location privacy using secure multi-party computations and described the protocol for privacy preserving based on 3-way authentication and key agreement.

4.5.2 Privacy Preserving Dating System

As a simple example of online collaboration where privacy preserving could be seen as a natural requirement to the system, we can consider a privacy preserving dating system. In such systems, participants should be able to describe their interests and preferences weighting them by importance. We can look at such a system as a matchmaking process that matches participants by their interests and preferences. Privacy preserving in this context means that the data describing participants preferences will remain private that is not available to other participants or the system running the application. A possible approach to implementing such a system is to use private matching and base the private set intersection on it as described in [113].

4.5.3 Privacy Preserving Monitoring in Wireless Sensor Network

Generally, sensor networks support distributed interaction with the physical environment through measuring and aggregation of data in order to create a dynamic global view. Various streams of measured data can be used to monitor and detect events of interest. Each event is represented as a set of values of monitored parameters.

Consider a wireless sensor network for monitoring vital sign parameters from patients in a metropolitan area. Such a network includes body sensors communicating with a receiver unit carried by a patient, which in turn can use another wireless hoop (for example, 3G telecommunication solution) to transfer data to a central base station. Sensor networks transmit monitoring data via a wireless medium and are thus vulnerable in terms of privacy and security. Sensor measurements represent private information about monitored objects, which requires that data transmissions and data flow within and out of the sensor network should be protected.

One approach to protect privacy in such sensor networks would be based on the idea that each sensor node delivers a part of the sensed data, called a share [254]. Each sensor share is a subset of monitored parameters assigned to that sensor. For example, we can define a function *Share* that maps individual sensors to a power set of monitored parameters. Thus, in order to obtain complete information about the monitored environment (status information), a base station should collect shares from all sensors in the network. The shares should be selected in such a way that individual sensor outputs are not sufficient for reconstructing complete status information. Intelligible reconstruction of the status information is only possible when a certain number N of distinct shares is available, where N is called an intelligibility threshold. For example, assuming that each share is associated with one monitored parameter for each sensor in a sensor network, complete status information includes knowledge of all sensor parameters. The complete status information is associated with the lowest security and privacy requirements, since it reveals all the data delivered by sensors.

4.5.4 Privacy Preserving Electronic Surveillance

In [115], the authors consider the problem of privacy preserving collection about individuals of data conditionally dependent on surveillance authorization. The authors consider how to monitor activities of only those individuals whose surveillance is authorized without disclosing the identities of monitored individuals to the data collecting entity, while ignoring individuals whose monitoring is not authorized.

The problem can be formalized as follows. It is assumed that the set of all identities U is a subset of the set S of identities for which monitoring is authorized. Let Alice be the monitoring agency that knows S, and Bob be a data-collecting entity that can collect data about activities of elements from U that he observes. The privacy preservation in this setting means that Alice can learn about activity of identity p from U if and only if p is also in S, but collecting entity Bob cannot learn anything about the membership of p in S. In [115], the authors propose privacy preserving solutions for this problem.

4.5.5 Privacy Preserving Credit Checking

A privacy preserving credit checking problem was considered in [114]. It deals with the process of applying for a loan that can be described as follows.

Assume that Bob wants to borrow money from a lender, Linda. Before giving a loan to Bob, Linda checks Bob's credit history to find out whether Bob is trustworthy and capable of repaying the loan. To do a credit check on Bob, Linda requests a credit report about Bob from a Credit Report Agency. Linda determines if Bob qualifies for the requested loan based on her qualification conditions. As we can see from this description, some private information will have to be revealed during this process (if it is done in a traditional way). For example, Bob's financial information such as borrowing history, spending habits, etc., are commonly described in the credit report. Meanwhile, Linda does not need to know all information from the credit report. What she really needs is to check whether her qualification conditions are satisfied. However, in some cases these conditions can also be private and Linda will not reveal them to anybody including a Credit Report Agency. Thus, the privacy preserving solution of credit checking would involve approving a loan application as it is described above where both the borrower's (Bob) private information from the credit report and the lender's (Linda) qualification criteria remain private.

Formally, the problem is defined as follows. We assume that a credit report is presented as a set of attributes $a_1, a_2, ..., a_m$ where a_i is either Boolean or an integer, and the qualification criteria $c_1, c_2, ..., c_n$ where each criterion c_i is a function on a subset of the attributes. The lender's qualification policy that determines whether a borrower qualifies for a specific loan is defined based on what combination of criteria $c_1, c_2, ..., c_n$ are satisfied on $a_1, a_2, ..., a_m$. The proposed solution utilizes secure multi-party computations that solve this problem efficiently. It requires communication and computation resources proportional to the size of the credit report and lender's policy.

4.6 Summary

In this chapter, we have provided a brief overview of ideas behind the use of secure multi-party computations in privacy preserving applications. By using simple examples, we demonstrated how SMC techniques have been used to develop novel applications with privacy preservation as an essential property. However, the most common drawback of SMC protocols, which substantially impacts their applicability, is their inefficiency. They require both considerable computational and communication resources. Not many, if any, of such applications are implemented. However, as both the availability of such resources and privacy concerns are continuously growing, one would expect that many such applications may be implemented in the near future (one such example is described in [43]). An important problem is whether practical solutions exist that are based on an ideal security model. Therefore, future research should seek other security models that can provide low-cost practical solutions with acceptable security level for given type of applications [85]. Finding such practical solutions that balance security and efficiency is an important research area.

5

Privacy and Data Mining in Telecommunications

5.1 Background and Context

Generally, data mining is the process of extracting knowledge from data and can be used to support knowledge-based decision making (for example, to optimize performance, increase customer satisfaction, increase revenue, cut costs, etc.). There are several reasons to adopt data mining in telecommunications: competitive markets, high churn rates and huge data collections available [54]. Telecommunications industry is considered as one of the most data-intensive industries in the world and has adopted data mining very early.

Traditionally, telecom databases have contained call detail data, customer data, network data. Call detail data include average call duration, no-answer-calls, average number of calls per day, calls to/from a different area etc. Network data contains data generated by network components and are needed for efficient network management (for example, for fault isolation). Customer data contain information about company's customers such as names, addresses, billing information, payment history, etc. With these data, many companies in the industry have started to apply data mining for better customer analysis, marketing, fraud detection, network fault isolation and prediction [249]. Databases maintained by telecommunication companies are one of the biggest in the world and a big portion of data collected in these databases are potentially privacy sensitive especially because of the growing use of smart phones and associated privacy sensitive services (for example, spatial-aware). For example, AT&T maintains databases containing (in 2010) 323 terabytes of information.

Availability of non-expensive computational resources creates new opportunities for companies by collecting and analyzing (in real time) customer data together with data reflecting customers' behavior by monitoring use of personal mobile devices. The results of such analysis can be used to improve users' security and safety, and quality of life. It also can be used to optimize companies' performance and increase revenue by, for instance, better capacity planning. For example, social network analysis can be used to identify groups of potential churners and discovering association rules that can be used to identify the next best product. However, the data also can be used to violate users' privacy by analyzing users' habits, monitoring of users' activities and even for surveillance.

One possible way to prevent privacy threats and still be able to extract useful knowledge from the huge amount of available data is to use privacy preserving data mining and knowledge discovery [241, 243]. The main reason for the rapid development of privacy preserving is the growing awareness of the accumulation of a huge amount data (including spatial data) that may comprise a threat to the privacy of users. Such data can for instance reveal driving habits, shopping patterns, etc.

In this chapter we give an overview of selected approaches in the area of privacy preserving data mining with special focus on approaches that can potentially be used to develop privacy preserving applications in the area of telecommunication.

5.2 Motivation

Privacy preserving data mining is an active research area in data mining and knowledge discovery [241]. The main reason for the rapid development of this research area is the growing awareness that accumulation of a huge amount of easily available data on the Internet and availability of non-expensive computational resources comprise a threat to the privacy of users when, for example, personal identifiers such as addresses, names, etc., can be connected with other person related information [231]. For example, using data mining techniques, shopping or travel habits could, for instance, be extracted from the traces of information that may remain after Internet use, which in turn may be related to a person's identity, potentially revealing sensitive information. On the other hand, if privacy concerns are addressed properly, society may benefit from the knowledge that can be distilled from sensitive data. Different approaches have been developed to tackle this dilemma. One approach is to use data obfuscation that modifies original personal data in order to protect a person's privacy. Another approach is to develop privacy preserving data mining algorithms that protect a person's privacy in the sense that no private data is revealed, for instance, by using secure multi-party computations [212].

In this chapter, we give an overview of use of both approaches in the area of privacy preserving data mining with special focus on approaches that can potentially be applied in the area of telecommunication.

5.3 Data mining

Generally, data mining concerns the extraction of knowledge from potentially large collections of unstructured and structured data, such as medical records, telecommunication customers calling data, web discussion forums, etc. Basically, extracted knowledge consists of discovered patterns and associations that are hidden in the data. In this sense, data mining can be said to add new meaning to data. Searching patient data from all hospitals in the world for patterns could, for instance, uncover new relations between potential treatments and outcomes, symptoms and diseases, and so on.

A number of distinct approaches to data mining have been identified, namely, classification, association rule learning, clustering, multidimensional scaling [129]. These different techniques can be summarized as follows:

- Classification techniques support categorization of elements within a data set into predefined categories. For example, customer classification in the telecom industry can be used to identify which value added services would be successful with a given customer segment that can be used to provide a flexible revenue model [39].
- Association rule learning concerns the discovery of elements that co-occur frequently within a data set. In telecommunications, association rule learning can, for example, be used on calling detail data to identify pairs of customers that frequently call each other (which in turn can be used to identify the so-called calling circles) [249]. Learning association rules from alarm databases of telecommunication networks are useful in locating problems in the network and filtering redundant alarms [251].
- Clustering techniques partition a data set into subsets so that the elements of each subset share common traits. Mobile service providers use these techniques to predict possible churners and take proactive actions to retain valuable customers. Clustering techniques can be used to build such predictive models for telecom churn [138].
- Multidimensional scaling techniques are used to detect meaningful underlying dimensions in a high-dimensional data set. It is a set of related statistical techniques that are often used for information visualization for exploring similarities or dissimilarities in data. For example, it can

be used for detection of anomalies such as abnormal customer traffic, abnormal call duration, etc. by determining whether the examined data belong to a normal or anomalous behavior (based on previous observations) [64].

5.4 Privacy Concerns and Data Mining

In practice, many potentially beneficial applications of data mining are constrained by privacy concerns. Since the aim of data mining is to uncover hidden patterns and correlations that are not explicitly given, it also has potential for uncovering sensitive information (patterns and correlations) that concerned parties consider to be private. As discussed in [249], when, for instance, the telecommunication company MCI launched its friends and family campaign in 1991, customers had to report their calling circles themselves in order to benefit from the campaign. From a marketing perspective, it would arguably have been more effective to proactively offer calling circle deals, using calling circles identified automatically by means of data mining on call detail data. However, uncovering such calling circles automatically, and using them for marketing purposes, could anger customers. Automatically identified calling circles can, for instance, reveal sensitive data about the customers.

Another example is using data mining techniques which may require data mining data distributed among different private databases. For example, to detect disease outbreaks, which may require data on disease incidents, patient background (electronic medical records database), and so on [252]. Since such data are sensitive, legal concerns may hinder their free use, even for beneficial purposes. In other cases, the target data may be partitioned across several organizations, thus organizational policy- and commercial concerns may restrict data use, in addition to legal ones.

Choosing between privacy and the benefits of data mining seems unavoidable. However, as it was shown in recent research, there are cases when they can be combined. The challenge lies in getting valid data mining results without learning the underlying data values. As we shall see in the following, previous statistical work on data disclosure and recent cryptographic techniques form the basis for current solutions.

5.4.1 Data model

In considering privacy preserving data mining, we should consider two separate cases: either all the data are available in one repository or the data are distributed among different (private) repositories.

When all the data are available in one repository, the techniques combining de-identification and perturbation of information should be utilized.

When all the data are distributed among different private databases, it is important (from the point of designing of data mining techniques) to consider how data are partitioned among the involved parties. When organizations collect the same kind of data about different entities (e.g., people, traffic, etc.), we say that the data are partitioned horizontally (from a database perspective), i.e., the same schema is used to store the data at each site. When organizations collect different kinds of data, perhaps on the same entities, we say that the data is partitioned vertically (from a database perspective) and organize data using different schemas.

For example, assume that n telecommunication companies want to cooperate on fraud detection (detecting misuse of mobile phones). Each company maintains a database with daily aggregated customer calling data including the number of international calls associated with each customer per day. In addition, the law enforcement unit maintains a database on previous known frauds (it can also be maintained by companies all together). Because of legal restrictions and/or business/privacy reasons none of the companies want to reveal their databases to each other. However, they want to construct rules for detecting frauds using the joint data. Since calling data databases contain data on own customers, they can be seen as horizontal partitioning of data. Previous fraud databases contain data that are different from calling data and may contain overlapping set of customers, and can be seen as vertical partitioning (different schema, same entities).

5.4.2 Main techniques

We consider two main approaches for achieving privacy preserving data mining: the first approach is based on data perturbation, and the second one is based on secure multi-party computations. We will briefly overview these classes before we discuss specific privacy preserving techniques for classification, association rule learning, clustering, and multi-dimensional scaling.

Data perturbation modifies sensitive data by swapping and adding randomized noise so that it loses its sensitive meaning, but still retains statistical properties of interest. The idea of using perturbation is based on the fact that most data mining methods generalize data. By performing perturbation in the way that preserves statistically significant properties of data may be useful when the owners of the data elements need to protect their data. Privacy can be preserved since only randomized and transformed data will be revealed even when all the data are available in one repository.

Secure multi-party computations are used when the data are distributed among different (private) repositories. It is assumed that at the end of the computations, the parties should only have learned the result of the computation (in addition to the input the party itself provided to the computation). Secure multi-party computation methods used for privacy preserving data mining are typically based on such building blocks as secure sum, secure set union, secure size of set union etc [212]. For example, secure sum protocol describes how to calculate the sum of distributed items without revealing their true values [68]. (See also Chapter 4 on secure multi-party computations in this book.)

5.4.3 Privacy Preserving Classification Techniques

Constructing a classifier typically involves so-called training. In brief, training means using already categorized data elements, called training data, to derive a procedure for classifying new and previously unseen data elements. For instance, one might want to learn a procedure for predicting fraud attempts based on the number of international calls of customers, using the data from different private databases as training data. One approach could be estimating the parameters of a parametric model, so that the model predicts as accurate as possible (e.g., using a maximum likelihood based approach). Thus, privacy preserving classification techniques essentially address two questions:

1. How can we train a classifier without revealing the training data itself?
2. How can we classify new unseen data elements without revealing those data elements?

It turns out that in most cases each type of classification technique requires a tailored solution, targeting either vertically or horizontally partitioned data, and using data transformation/randomization and/or secure multi-party computations. For instance, in Privacy Preserving Nearest Neighbor Search for horizontally partitioned data [252], the goal is to find the training data element that is nearest to the data element that we want to classify; however, without revealing any of the data elements, only the final classification.

Similarly, a Naive Bayesian classifier for vertically partitioned databases may need to be trained when not all of the parties know the class attribute of the training data [241]. Thus, in the worst case, one party knows the class attribute of the training data, while the other parties know the other attributes. In brief, the problem is that training a classifier requires data element attributes and class information, yet privacy concerns may not allow this information to be shared. privacy preserving Naive Bayes classifiers on both vertically as well as horizontally partitioned data are analyzed in [243].

5.4.4 Privacy Preserving Association Rule Learning

The goal of association rule learning is to find specific patterns that represent knowledge in generalized form without referring to particular data items. Because of this one might say that association rule learning only represents an indirect threat to privacy. However, traditional methods require access to the data set in order to be able to find association rules.

As we described above, we assume that the databases are owned by different parties and no one wants to disclose their data to other parties. The main concern is how to avoid revealing data to other parties. Both data transformation/randomization and secure multi-party computation techniques have been applied to develop privacy preserving methods for association rule learning.

In the first approach [105], data are randomized such that the true values cannot be estimated with sufficient precision. The typical problem can be formulated as following. Assume that there are several customers having databases containing private information and one server which are interested in learning association rules based on statistically significant properties of this distributed information. Customers protect the privacy of their data by perturbing data with some randomization algorithm and then submit the randomized version of data to the server. The framework for mining association rules from data that have been randomized to preserve privacy is described in [105].

The second approach assumes that data are distributed between two or more sites and the purpose is to learn global association rules without revealing data to other sites. This approach was applied to both horizontally and vertically partitioned data [161, 242, 243].

5.4.5 Privacy preserving clustering techniques

The goal in clustering is to partition data elements into clusters so that the similarity among elements belonging to the same clusters is high, and so that the similarity among objects from different clusters is low. In privacy preserving clustering, a main goal is to find the clusters in the data, without revealing the content of the data elements itself. For instance, two or more companies may decide that performing clustering on customer data might improve their direct marketing. However, they are not willing to reveal their own customer data to the other party. Again, note that the data may be partitioned vertically and/or horizontally among the involved parties. To exemplify, in [68] a scheme for privacy preserving of the so-called EM-clustering is proposed that only reveals aggregated quantities using multi-party secure computations. In [204], data transformation/aggregation is used, so that the clustering performed on the distorted data is still valid.

5.4.6 Multidimensional scaling techniques

Informally, Multidimensional Scaling (MDS) is the process of transforming a set of points in a high dimensional space to a lower dimensional one while preserving the relative distances between pairs of points. This property is important in the context of data visualization where it is important to preserve relative relationship between data items while reducing dimensionality. The privacy preserving visualization problem for this case can be formulated as following. Assume that Alice and Bob have two sets of private data items that represent sets of points in m-dimensional Euclidean space. They wish to visualize jointly both sets without revealing data items where none of visualized points can be correlated with items in the original data sets and therefore it is not possible to find the origin of any point. In [29], authors propose using the non-metric MDS technique to preserve data utility for clustering whilst maintaining privacy. The authors demonstrate how non-metric MDS can be used as a perturbation tool such that the perturbed data can be used in clustering analysis without compromising privacy.

5.5 Summary

Privacy preserving data mining is about how to develop models without seeing the data, that is, how to produce accurate mining results without disclosing "private" information. The area of privacy preserving data mining encompasses a number of novel and challenging problems that spans several fields. It is particularly important for telecommunication industry where companies maintain huge databases of privacy sensitive (almost real-time) information about individuals. It is a way to increase the use of this information for benefits of the industry and society.

6

Requirements for Cellular Systems Subscriber Privacy

6.1 Setting the Scene

Cellular, or mobile, systems are covered in Chapters 6 – 8. In this chapter we focus on generic security- and privacy requirements for the cellular/mobile systems. In Chapter 7 we continue with a brief analysis of the GSM/GPRS, UMTS and LTE/LTE-Advanced systems and, in Chapter 8, we take a look at ways to improve the subscriber privacy beyond current state.

6.1.1 System Perspectives on Subscriber Privacy

The purpose of this chapter is to examine and describe access security and subscriber privacy in cellular systems. The main focus is on subscriber privacy, and here it refers mostly to data privacy, identity privacy, location privacy and untracebility. We emphasize those aspects which are particular to the mobile systems context or on privacy aspects which become acute and urgent in mobile setting. First, we start out by outlining the main cellular system architecture. This architecture is the 3GPP system architecture, which encompasses the 2G system GSM, the 2.5G system GPRS (incl. EDGE), the 3G system UMTS and 4G system LTE/LTE-Advanced. We successively investigate GSM/GPRS, UMTS and LTE/LTE-Advanced with respect to access security and subscriber privacy. The 3GPP system architecture is a complex subject and we will only scratch the surface here. We dig a little deeper in the area of access security and subscriber privacy, but obviously our presentation cannot fully cover all the details. We encourage the interested reader to check out the following references for a more complete picture [134, 140, 167, 172–174, 196, 218]. This chapter draws on material from the following books and papers [168–170, 173, 177, 179].

6.1.2 World View

The most widely used systems today are GSM/GPRS, cdmaOne, UMTS and CDMA2000. The GSM/GPRS and UMTS systems belong to the so-called 3GPP family of systems and the cdmaOne and CDMA2000 belong to the 3GPP2 family of systems. The 3GPP and the 3GPP2 are standards defining bodies, but neither 3GPP nor 3GPP2 actually ratifies the systems.

The emerging 4G systems are the first truly All-IP Network (AIPN) mobile architectures. Good old circuit-switched speech is gone. There are two main alternative RAN implementations for IMT-Advanced, the 3GPP based Long-Term Evolution (LTE) and the IEEE defined WiMAX systems (IEEE 802.16). However, neither system formally qualified as a 4G system according to the IMT-Advanced requirements, and they are now often termed as 3.9G systems. So IEEE and 3GPP had to come up with enhanced versions of their systems to qualify, and thus was born LTE-Advanced ("Release 10") and WiMAX 2 (IEEE 802.16m). System wise the difference between the 3.9G systems and the 4G systems is almost completely isolated to the radio access part.

The description given in this chapter is for the 3GPP-based systems. The 3GPP2 systems are very similar, but there are some differences. An overview over the security differences for 3G systems can be found in [218].

6.1.3 Notes on Nomenclature

We use the terms "cellular" and "mobile" interchangeably in this chapter. The security architecture for LTE and LTE-Advanced is exactly the same. The channel aggregation in LTE-Advanced allows for higher bandwidths, but the access security architecture remains the same. Whenever we indicate LTE it shall be understood to be applicable to both LTE and LTE-Advanced.

6.2 Privacy Threats in Mobile Systems

6.2.1 Primary Threats, Associated Threats and Compound Threats

In Sec.1.4 we identified a set of privacy aspects, including:

1. Data Privacy
2. Identity Privacy
3. Location Privacy
4. Traffic Privacy
5. Movement Privacy / Untraceability

6. Transaction Privacy

We first take a brief look at the first three aspects, with Identity- and Location threats being discussed together. We will also briefly investigate the last three aspects, which we consider to be associated or compound threats. However, first we need to define the principal parties in the mobile context.

6.2.2 The Principal Parties

The principal parties in a mobile system are the subscriber, the visited network and the home network. The subscriber is represented by his/her User Equipment (UE), which amongst other things include the Subscriber Identity Module (SIM). Two types of SIM cards are used in the 3GPP systems, the GSM SIM card (GSM/GPRS) and the UICC card (UMTS/LTE). The UICC will run a USIM software module, while the GSM SIM includes similar software embedded on the device. A UICC card is shown in Figure 6.1.

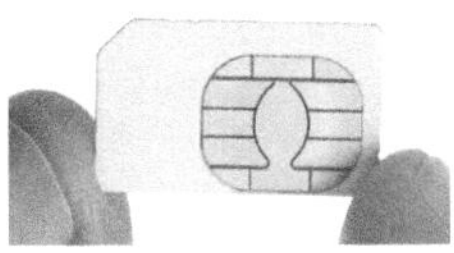

Figure 6.1 The UICC

- *The Subscriber (user)*
  The subscriber is the party who has a mobile subscription. This entitles him/her to mobile services at "home" network (HPLMN). In technical terms, the subscriber is represented by his/her SIM and is identified by an international mobile subscriber identity (IMSI).

- *The Home Public Land Mobile Network (HPLMN)*
  The HPLMN is the party with which a subscriber has a subscription. It will have subscriber databases (HLR/HSS) and gateways (GGSN/PGW), and commonly have radio access network (RANs).

- *The Visited Public Land Mobile Network (VPLMN)*
  The VPLMN is the party which has the radio access networks (RANs). The VPLMN and HPLMN have signed *roaming agreements*, which allow HPLMN subscribers to roam onto the VPLMN radio networks.

- *External Parties*
  The external party is any other entity. This includes threat agents, adversaries and intruders. It may also include regulatory bodies and law enforcement agencies (LEA), in addition to third party service providers.

The terms HPLMN and VPLMN are relative to the subscriber. Figure 6.2 shows the LTE roaming architecture, with the HPLMN and VPLMN interfaces. The signaling procedures are essentially the same for a subscriber whether he/she is attached to a RAN belonging to the HPLMN or to a VPLMN. The terms "home network" and "visited network" will typically be used to denote RANs belonging to the HPLMN and VPLMN, respectively.

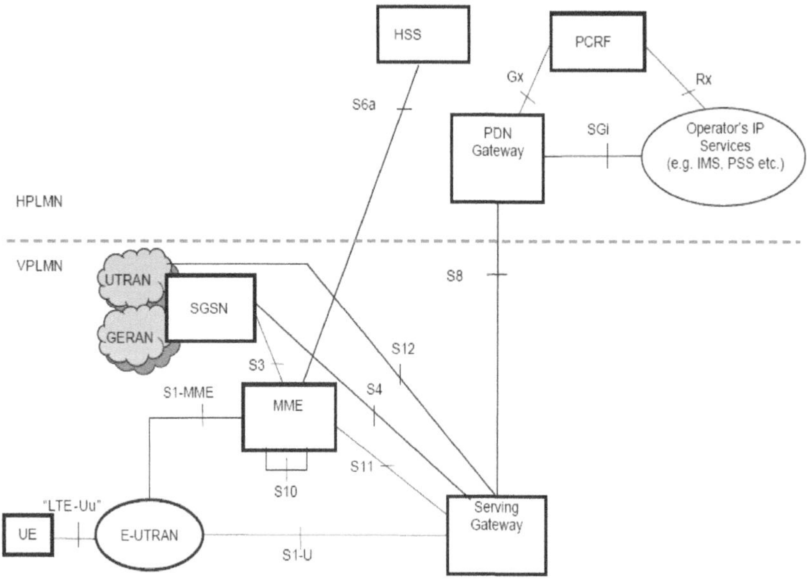

Figure 6.2 The LTE Roaming Architecture (TS 23.401 - Figure 4.2.2-1))

6.2.3 Data Privacy

Data privacy is a generic threat and is, in principle, no different from data privacy for fixed line connections. However, a wireless connection is different from a fixed line connection in the way it is exposed. Radio signals propagate and any receiver in the coverage area may read the signals. Radio transmissions may be directional and the coverage area may be small, but a radio connection is anyhow far more exposed than a fixed line connection.

Over-the-air connections are an easy target for eavesdropping. Every mobile device within the radio range of the signals can potentially listen-in on the communication. Well behaved mobile device will not accidentally or otherwise listen-in on conversations, but technologically it is not at all difficult to read the radio signals. The only realistic defense is cryptographic data confidentiality protection (encryption).

Data privacy is also a concern for the data stored on the mobile device. Once again the fact that a mobile device is mobile means that its more exposed. This time the exposure is towards theft and towards the owner misplacing/losing the device. *Apps* that steal confidential information is another threat to data privacy. This type of malware already exists, predominantly trying to steal identity related data and credit card numbers (phishing attacks etc.). This is an expected trend. We store more and more valuable data on the mobile device and we use it for more purposes; Both the exposure and the value of the assets increase, thus making it a preferred platform for attacks.

6.2.4 Identity- and Location Privacy

The main privacy issues identified for cellular systems have to do with the mobile nature and the fact that the connections are over-the-air. The subscriber position isn't fixed anymore and that makes location an interesting attribute. The location attribute alone can only reveal a part of the picture, but if one knows the whereabouts of specific people then the location information attribute becomes much more privacy sensitive.

A typical mobile device is often more-or-less permanently turned on. The mobility management software in the device will periodically confirm its location to the network and it will contact the network when the location area/roaming area changes. Many smart-phone services and apps will also periodically contact their servers to keep sessions alive. Identity information should obviously not be revealed in plaintext over-the-air, but as it turns out it is quite difficult to avoid revealing the subscriber identity. It is also hard not to reveal information, that taken together, allows an intruder to infer the identity of the subscriber. Another aspect of *location* is that a geographically distributed intruder, with two or more receivers within the range of the radio signals, will necessarily be carrying out triangulation on the signal. We haven't explicitly mentioned it, but it is understood that the intruder also knows the time of the observations. The intruder observation may then potentially be $<time, location, identifier>$. The $identifier$ in question can be any statistically unique collection of observed data that serves to identify the entity (an *emergent* identity).

Advanced triangulation and location determination are inherent features of a cellular system. The E911/E112 (Emergency Call) service, which is mandatory in most jurisdictions, requires that the operator be able to determine the approximate location of the caller. The EC service requirements only applies to emergency calls, but one also has generic location services and Lawful Interception (LI). These services use an array of different location determination techniques depending on the specifics of the radio system, including mobile-assisted use of the GPS receivers which many mobile devices have. The location-based services may be seen as a huge privacy threat, but we here assume that use of these services require positive user consent. Then technically we no longer consider it a privacy threat. Of course, one shouldn't be too naive here; agreements may be abused and trust may be betrayed, but that is outside the scope of this chapter.

6.2.4.1 Traffic Privacy

Traffic privacy is concerned with the pattern of exchanged data. This aspect is not specific to a cellular system, but in a cellular system a privacy intruder may be able to collect location data and correlate it with the traffic pattern. Such an intruder can either be stationary, which makes for a local threat, or it may be mobile itself and may potentially follow entities as they move around. Capable intruders may also have a geographically distributed network of devices. In some contexts one may also consider the cellular network to be a threat agent, and then threat is ever-present. Protection against traffic privacy invasion is difficult and in a cellular system it is very difficult to hide the location of a node. There are ways of masking out data transmission patterns, but these generally only work if all mobile units engage in the same traffic pattern. This is highly impractical, very costly and it is consequently not done for any public cellular system. And, not only is it impractical, but it would also be counter productive with respect to subscriber privacy. A steady traffic pattern is very easily tracked and even if it would be difficult to determine the identity or type of traffic per se, in the long run the the actual movements would more-or-less inevitably betray the identity of the mobile device.

6.2.4.2 Movement Privacy / Untraceability

Movement privacy is your right not to divulge your movements. Movement privacy is related to location privacy, but instead of applying to a specific observation it applies to a time series of observations. That is, the privacy intruder is able to obtain a time series of the $<time, location, identifier>$ tuple. To obtain a time series, or to track an entity, requires as a minimum the ability to distinguish the tracked entity from other entities. Thus, the privacy intruder must be able to identify the tracked entity. This does not mean that the intruder knows the "real" identity, but he/she need at least some kind of distinguishing identifier to be able to track the user. This "emergent" identifier can be just as real as any properly designated identifier, and over time the intruder may be able to associate it with other identifiers.

6.2.4.3 Transaction Privacy

Transaction privacy is the right not to divulge transactions. The concept is broad and could include any transaction that you would want to keep private.

Of course, in a cellular system the intruder may already be able to track the target subscriber. One must also assume that the intruder is able to detect communication patterns, maybe to the extent that the intruder knows the caller/callee information and/or termination points for data connections.

When knowledge of a transaction becomes valuable or privacy sensitive depends a lot on the "target", but also on the intruder. The mobile network cannot possibly be responsible for the usage patterns or the mobility patterns of the subscribers. This makes it next to impossible for the network to directly protect your transaction privacy; it can at best help you to protect your network identity and your location.

6.3 Privacy Requirements

6.3.1 The 3GPP Systems

The 3GPP is the body responsible for the GSM, GPRS, UMTS and LTE systems. The access procedures and access security schemes in the 3GPP systems have quite a lot in common, including the general identity presentation, the location registration and the authentication and key agreement signaling procedures. The details do indeed vary between the different systems, but the general schemes are the same. So, it is not surprising then that the privacy issues are more or less the same with all these systems. The 3GPP security architecture for 3G (UMTS) defines a set of subscriber privacy requirements [10], which we shall briefly outline in the next subsections. The 3G requirements also apply to the 4G systems (LTE), and the 3GPP SA3 (Security) working group now formally has both security and privacy included in the terms of reference (`www.3gpp.org/SA3-Security`).

The 3GPP2 requirements (cdmaOne/CDMA2000) are basically the same as for 3GPP. Indeed, the security architectures are almost identical and the systems even share the basic UMTS AKA algorithm [167, 173, 218].

6.3.2 User identity confidentiality

User identity confidentiality (UIC) is recognized by 3GPP and it is related to the globally unique International Mobile Subscriber Identity (IMSI). Figure 6.3 depicts the IMSI and the associated TMSI identifier.

UIC in this context is defined as "*the property that the permanent user identity (IMSI) of a user to whom a services is delivered cannot be eavesdropped on the radio access link*" [10].

The scope is limited and confined to the radio link. This was more-or-less sufficient in 1999, when the requirement was framed for UMTS (3GPP Release 99), but in an All-IP Network world it appears dated. The IP-backbone in the access network is, of course, not an open network, but an unprotected IP network is easy to eavesdrop on (it is not mandatory to protect the backbone). So we may consider the requirement necessary, but it is by itself not sufficient.

6.3.3 User location confidentiality

User location confidentiality is described by the 3GPP as "*the property that the presence or the arrival of a user in a certain area cannot be determined by eavesdropping on the radio access link*" [10]. It is understood that this applies to an *identified* user and that it refers to the above mentioned IMSI/TMSI identifiers. This requirement is also *necessary*, and it may be *sufficient* as well. We must qualify the "sufficient" part to include a condition that location data isn't associated with the identity in the (unprotected) signalling data in the radio access network backbone.

6.3.4 User untraceability

User untraceability is defined by the 3GPP as "*the property that an intruder cannot deduce whether different services are delivered to the same user by eavesdropping on the radio access link*" [10].

The definition is slightly different from a direct requirement on protection against tracking in that it includes the "different services" formulation. Thus, there is an element of traffic analysis protection and a measure of transaction privacy baked into the definition. The requirement itself is defined in relation to the "same user". This requirement too is necessary, and it appears to be sufficient. That is, we still have the precondition that we had for user location confidentiality.

6.3.5 Traffic- and Transaction Privacy

The 3GPP does not have any explicit requirements for traffic privacy in the access security area, but the Network Domain Security (NDS) contains some provisions for traffic privacy. That is, the NDS/IP specification [13], which provides an IPsec profile for use with 3GPP network, takes advantage of the traffic privacy that can be provided by IPsec. This is not very much, but then there really wasn't a requirement for traffic privacy in the first place.

Transaction privacy is not addressed directly in the 3GPP specifications, but if "user untraceability" is successfully addressed then one may consider the preconditions to be fulfilled.

6.4 Identity- and Location Privacy in Mobile Networks

6.4.1 The Main 3GPP Subscriber Identifiers

The main subscriber identifiers are the permanent IMSI identifier and the temporary identifiers (TMSI/GUTI). The internal procedures *location updating/registration*, *handover* and *paging* use the IMSI and/or the TMSI/GUTI. The authentication and key agreement (AKA) procedures are also based on the IMSI. The phone number, in accordance with the ITU-T E.164 ISDN numbering scheme [148], is not a primary identifier in the 3GPP systems.

Another difference between the phone number (MSISDN) and the IMSI is that the phone number is normally a public identifier. The IMSI, while not secret per se, is not public and it is generally not known to the user.

6.4.2 The Permanent Subscriber Identity

The IMSI is the primary subscriber identity in all IMT and IMT-Advanced systems. It is *the* primary identity and it is associated 1-to-1 with the subscriber security credentials. The IMSI, the associated security credentials and the authentication algorithms are all stored on the tamper resistant subscriber identity module (SIM). The SIM is a smart card and may either be a GSM SIM type or a UICC/USIM type (UMTS and LTE). The SIM cards comes in a number of different form factors (FF).

The structure of IMSI is defined in the ITU-T E.212 recommendation [149]. The normative 3GPP reference is TS 23.003 "Numbering, addressing and identification" [4]. The IMSI is a globally unique system internal identifier, consisting of a mobile country code (MCC), a mobile network code (MNC) and a mobile subscription identification number (MSIN).

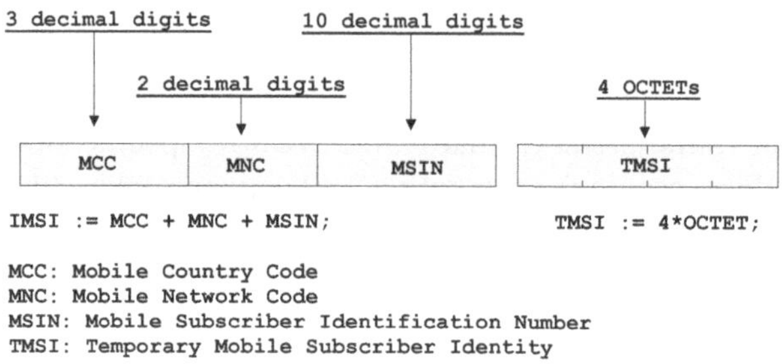

Figure 6.3 The IMSI and TMSI identifiers

Figure 6.3 depicts the "classic" layout of the IMSI. The MNC *may* also have 3 digits, in which case the MSIN must be reduced in length. The length of MSIN should be interpreted as having "at most" the said length. The actual length of MNC and MSIN is decided by the national regulator.

6.4.3 The Temporary Mobile Subscriber Identity

The Temporary Mobile Subscriber Identity (TMSI) used in GSM/GPRS and UMTS is a temporary replacement for IMSI. The TMSI is assigned locally by the VLR/SGSN and the TMSI is only valid within the associated area. There is no formal requirement to restrict the lifetime of the TMSI, but in principle it may be renewed whenever the mobile device and network exchanges control information. So, what is the purpose of the TMSI? If we dig into the requirements for identity- and location confidentiality, we find the following (TS 33.102 [10]):

> To achieve these objectives, the user is normally identified by a temporary identity by which he is known by the visited serving network. To avoid user traceability, which may lead to the compromise of user identity confidentiality, the user should not be identified for a long period by means of the same temporary identity. To achieve these security features, in addition it is required that any signaling or user data that might reveal the user's identity is ciphered on the radio access link.

> Clause 6.1 describes a mechanism that allows a user to be identified on the radio path by means of a temporary identity by which he is known in the visited serving network. This mechanism should normally be used to identify a user on the radio path in location update requests, service requests, detach requests, connection re-establishment requests, etc.

So what we have is that User identity confidentiality, User location confidentiality and User untraceability all rely on the TMSI scheme. The TMSI is itself a 32-bit unstructured bit field (read: an unsigned 32-bit integer). For the TMSI scheme to be successful, it is not sufficient that it is assigned in encrypted mode and used in plaintext. It is also essential that the there is no apparent pattern to the assigned TMSI values. In particular, the assignment must not be sequential or otherwise contain structure or information which may be associated with the subscriber.

6.4.4 The Globally Unique Temporary Identity

The Globally Unique Temporary Identity (GUTI) is defined in TS 23.003 [4] and is used in LTE. It serves much the same purpose as does the TMSI. Functionally, the GUTI is a superset of the TMSI. The GUTI is defined as:

$$GUTI := GUMMEI || M{-}TMSI$$

The Globally Unique MME Identifier (GUMMEI) is constructed from the MCC, MNC and MMEI. The MCC and MNC have the same format as one finds in the IMSI.

$$GUMMEI := MCC || MNC || MMEI$$

The MME Identifier (MMEI) is divided into two subfields, the MME Group ID (MMEGI) and a MME Code (MMEC). The MMEGI is two octets wide and the MMEC is one octets wide.

$$MMEI := MMEGI || MMEC$$

The M-TMSI is a 32-bit wide "local temporary UE identifier" and the M-TMSI format is *exactly* identical to the TMSI used in GSM/UMTS. The M-TMSI alone is only valid within the area of the MME that assigned it.

$$S{-}TMSI := MMEC || M{-}TMSI$$

In 2G/3G, the TMSI is assigned by the VLR/SGSN and is only valid within the VLR/SGSN service area. Paging in 2G/3G takes place within that area and the TMSI can, therefore, be used directly as a paging identifier. In LTE, the paging is done for the whole MME Group and the M-TMSI cannot be used for paging alone. To solve this problem one uses the S-TMSI for paging purposes, which in addition to the M-TMSI also includes the MME code. The S-TMSI is unique within the MME group. To maintain credible subscriber privacy the M-TMSI must be frequently renewed.

6.4.5 Identifiers and Privacy

The following internal identifiers are considered to be the most important ones for identity- and location privacy.

- **IMSI** - Permanent/primary identity, all generations.
- **TMSI** - Local validity. No formal restriction on lifetime, only 2G/3G.
- **GUTI** - Global validity. No formal restriction on lifetime, only 4G.
- **M-TMSI** - Local validity. No formal restriction on lifetime, only 4G.
- **S-TMSI** - Regional validity. No formal restriction on lifetime, only 4G.

The temporary identifiers (TMSI,GUTI,M-TMSI,S-TMSI) are assumed to be relatively short lived, but there are no formal requirements restricting the lifetimes in the security architecture [10, 16]. Also note that the "regional validity" could easily encompass a whole network in smaller countries.

The above list is also not exhaustive and there exists other identifiers and identifiable information elements, such as the tracking area identity (TAI) and the sequence number (SQN), which are used in the system and which may be used to track a subscriber. The SQN is used to keep track of the permissible security context credentials (AV/EPS-AV). We shall return to the SQN and see how it is masked to avoid tracking (Section 7.2).

6.5 Location Updating and Assignment of TMSI

The following is an outline of the *local updating (LU)* control plane procedure in GSM/GPRS. The principle is identical in UMTS and LTE, although the details vary. Figure 6.4 depicts an example sequence. In the LU procedure, encryption is used to conceal the TMSI assignment. This is how it should be, but be warned that encryption is an option. So, when we say that the TMSI will "always" be assigned in ciphertext form we are assuming that encryption is being used in the network in the first place.

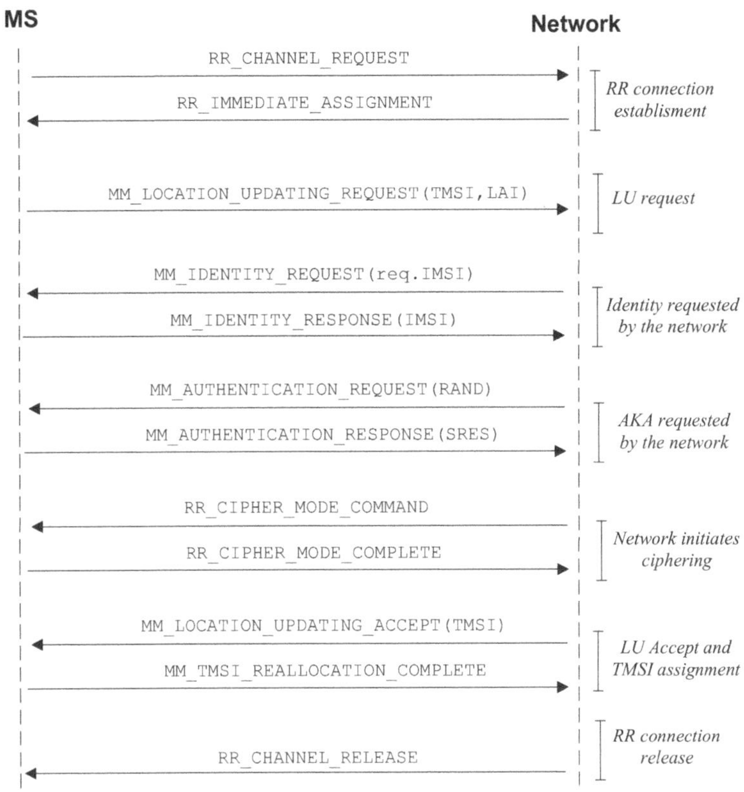

Figure 6.4 GSM Example – Local Location Updating

6.5.1 Local Location Updating and TMSI Assignment

The local location updating procedure in GSM/GPRS is part of the Mobility Management (MM) protocol and is specified as part of the DTAP protocol. The original DTAP specification is found in TS 04.08 [1], and it is Chapter 4.4 that is relevant here. Note that TS 04.08 is only valid for pre-UMTS versions of the 3GPP architecture, but rest assured that the principles for TMSI assignment still applies. The local location updating sequence may be triggered by four different events:

- MS power-on
- MS relocation to new Location Area
- Periodic updating (Timer T3212 expiry)
- VLR/MSC does not recognize the MS (during MM connection attempt)

The LU procedure is can only be executed in idle-mode; that is when the MS is not engaged in a call. The LU is also exclusively initiated by the MS. The LU procedure starts with the MS executing a channel request (`RR_CHANNEL_REQUEST` on an *access burst*). The network responds with assignment (`RR_IMMEDIATE_ASSIGMENT`) of a stand-alone dedicated control channel (SDCCH). The LU procedure is then executed over the SDCCH channel. Figure 6.4 shows an example of a successful location updating procedure. The figure is an amended version of Figure 7.9 in TS 04.08 [1].

The example outlines the LU procedure for cases where the VLR/SGSN area does not change. There will typically be several location areas within one VLR/SGSN area so this is a common occurrence. There is also a "global" LU procedure, which we will not go into here. Suffice to say that the global LU procedure is about updating the HLR/HSS when the subscriber moves onto a new VLR/SGSN area. This procedure may also involve cancelling LU information in the "old" VLR/SGSN area. The next sections explain the LU sequence example further.

6.5.1.1 The Identity Presentation

The MS will identify itself using TMSI or IMSI (*LU request*). This part of the sequence is in plaintext, and all identifiers used here will be exposed.

The inclusion of LAI (location area identity) in the example is necessary since the TMSI has only local validity. The same is achieved in LTE with use of the S-TMSI, which consist of the MME code and the M-TMSI. In the signaling outline we also see that the network side may request the IMSI directly. This is done when the network node (VLR/SGSN) does not recognize the {TMSI,LAI} tuple.

6.5.1.2 The GSM AKA Sequence

The network side may optionally decide to run the GSM AKA protocol. This will be done if the VLR/SGSN has no valid security context, which may happen when the subscriber is new in the VLR/SGSN area, or it may be done when the VLR/SGSN wants to renew the security context. In GSM/GPRS, one should ideally renew the security context for all LU events, but the actual policy is left as an operator decision. The GSM AKA procedure may also be triggered by other control plane sequences, like call setup etc.

6.5.1.3 The "Start Ciphering" Sequence

To start ciphering is an option, but it should really be done. That is because it will be the only way to confirm that the LU request was initiated by the claimed entity. Ciphering is also necessary if the VLR/SGSN wants to assign a new TMSI to the MS.

6.5.1.4 The TMSI Assignment

At the end of a successful LU sequence, the network will indicate that it has accepted the new location. This "accept" should always be verified request (read: subsequent to "start ciphering"). The network side may or may not assign a TMSI value at this time. In this case it is required that ciphering has commenced. The mobile device will respond to the "accept" before the LU procedure completes.

6.5.2 TMSI Allocation Summary

TMSI must *always* be assigned in cipher-text mode. This is essential for both security and privacy reasons. An external observer may be able to capture the plaintext IMSI and the encrypted TMSI. Subsequent to the assignment, the TMSI will be used for over-the-air identification purposes, and it will then always be used in plaintext. The use of TMSI will significantly reduce the IMSI exposure and the encrypted assignment will conceal any apparent link between the IMSI and the assigned TMSI. That is, an external observer should not be able to associate the IMSI and TMSI, and it should not be able to associate one TMSI with another TMSI used by the same mobile device.

6.6 Access Security in Mobile Networks

There is not one authentication and key agreement (AKA) scheme in the 3GPP systems, but rather one per generation. And, then there are variations between the schemes used in 3GPP and 3GPP2 too. However, the AKA schemes are all related and they all are fitted within a common access signaling structure which dictates the message exchange structure. That is, the message sequence is more or less fixed, but the content of the messages does indeed vary.

Technically, the AKA protocols are all based on variations of a basic challenge-response scheme, in which the two parties have a pre-shared long-term authentication secret. The procedure is roughly this: The network challenges the subscriber with a random challenge. Based on the challenge the subscriber computes a signed response. The "signing" is achieved by using a Message Authentication Code (MAC) function, which works under the control of a pre-shared secret key, and taking the challenge as the input.

The output from the MAC will be the signed response and associated key material. In the 2G scheme (the GSM AKA protocol) this is basically the whole procedure. In the 3G scheme (the UMTS AKA protocol) the challenge itself is authenticated, and in the 4G scheme (the EPS-AKA protocol) the derived key material is bounded to a specific serving network and the derived key material is only used for key deriving purposes (and not directly as a session key).

6.6.1 Security Contexts

The successful outcome of an AKA run is the establishment of an authenticated security context. The security context consists of, in addition to the corroborated identifiers, session key material. The key material is used, directly or indirectly, to maintain the established context. The key material may be used directly as session keys (GSM/GPRS and UMTS) or it may be used as key deriving keys (LTE). The derived keys in LTE are also part of the overall security context, although one here distinguishes between different types of security contexts. In either case, we have that secret key material belonging to the established security context is used for data confidentiality and data integrity purposes, and this will not only protect the communication but also maintain the security context.

At the network side, we have both the HPLMN and the VPLMN taking part in the AKA process. The HLR/AuC in the HPLMN is the location where the permanent security credentials are stored. After Release 5 the HLR/AuC is replaced with the HSS, which is a superset of the HLR/AuC. Derivation of the operative security context credentials also takes place in the HLR/AuC (HSS). The VPLMN is the network side party that carries out the challenge-response part of the AKA protocol. This will be the VLR/SGSN (2G/3G) or the MME (4G). The VPLMN will always be the party that initiates the AKA protocol and it will also be the party that initiates the over-the-air protection. The so-called over-the-air protection may extend into the core network (2G GPRS), it may be terminated in the RAN (UMTS) or it may be terminated at the base station (GSM/LTE). In LTE, the MME and the eNodeB will also be involved in deriving session keys from key deriving keys. At the subscriber side we have that the security context is created at the SIM. The SIM already contains the IMSI, the long-term pre-shared authentication secret (K) and the AKA algorithms. The SIM is also the termination point for the challenge-response procedure initiated by the VPLMN. That is, in LTE, things will be slightly more complicated, but this is anyway the main principle.

6.6.2 The Subscriber Identity Modules

There exists two different types of SIMs in the 3GPP system. In addition, the hardware may have different form factors.

- **GSM SIM**
  The GSM SIM is native SIM used in GSM/GPRS. It consists of both the physical smartcard and the logic functions on the card.

- **UICC/USIM**
  The UICC is the physical smartcard (Figure 6.1). The USIM is a secure and protected application running on the UICC. The USIM, running on a UICC, is the native SIM used in UMTS. The USIM is also the native SIM for LTE, but in LTE the USIM will be complemented by additional EPS AKA security functionality on the mobile equipment (ME).

Note:
The acronym "EPS" (Evolved Packet System) is used throughout in the security specification (TS 33.401 [16]. Like the acronym LTE (Long-Term Evolution) it is used to denote a 4G system or 4G functionality. Occasionally, one may also find the acronym EPC (Evolved Packet Core) used for more-or-less the same purpose.

6.6.3 Simplified Security Context Definitions

The `GSM Security Context` may be defined as:

$$GSM_{SC} = \{(IMSI, K_I, A3, A8, RAND) \rightarrow K_C|SRES\} \qquad (6.1)$$

The A3 and A8 are the AKA algorithms. In addition, there will also be the A5 cipher algorithm and an identifier called the $CKSN$ which is essentially a context identifier. The GSM Security Context is the native GSM and GPRS security context. It *may* be used when the subscriber accesses a UMTS network.

The `UMTS Security Context` may be defined as:

$$UMTS_{SC} = \{(IMSI, K, f) \rightarrow AV\},$$

where f is a set of UMTS AKA functions (f1,...,f5). The AV consists of the following: $AV = \{RAND, AUTN, RES, CK, IK\}$. We shall discuss the AV and the rest of the UMTS AKA parameters further in Sec.7.2. The UMTS Security Context is native to UMTS, but may also be used in

GSM/GPRS and in LTE. In GSM/GPRS, one then needs to convert the key material and in LTE one needs to map the context to an LTE context.

The `EPS Security Context` may be defined as:

$$EPS_{SC} = \{(IMSI, K, f, KDF) \rightarrow EPS{-}AV\}, \tag{6.2}$$

where f is a set of UMTS AKA functions (f1,...,f5). The KDF is a generic Key Derivation Function. The $EPS - AV$ consists of the following: $AV = \{RAND, AUTN, RES, K_{ASME}\}$. We shall discuss the $EPSAV$ and the rest of the EPS-AKA parameters further in Sec.7.3.

The `EPS Security Context` is native to LTE. It includes the UE-MME `NAS Security Context` (control plane) and it may include an `AS Security Context` (user plane). The `AS Security Context` only exists in a connected state. The `EPS Security Context` may be mapped down to a UMTS security context, and from there further down to a GSM security context.

7

The 3GPP Systems and Subscriber Privacy

7.1 Outline of the GSM Access Security Architecture

7.1.1 Background

The GSM AKA protocol was designed for the circuit-switched GSM system. It is also used for the GPRS system, which is basically an overlay system onto GSM for providing packet-switched services. The radio access network (RAN) is common to GSM and GPRS. The EDGE radio enhancements are also fitted onto the basic channel scheme provided in GSM, albeit at somewhat higher bit rates using an enhanced modulation method. The GSM/EDGE radio access network is called GERAN (Gsm/Edge RAN). The GSM AKA protocol may be used when accessing UMTS networks (UTRAN), but it is not permitted for access to LTE (E-UTRAN). Figure 7.1 shows the GSM access security architecture.

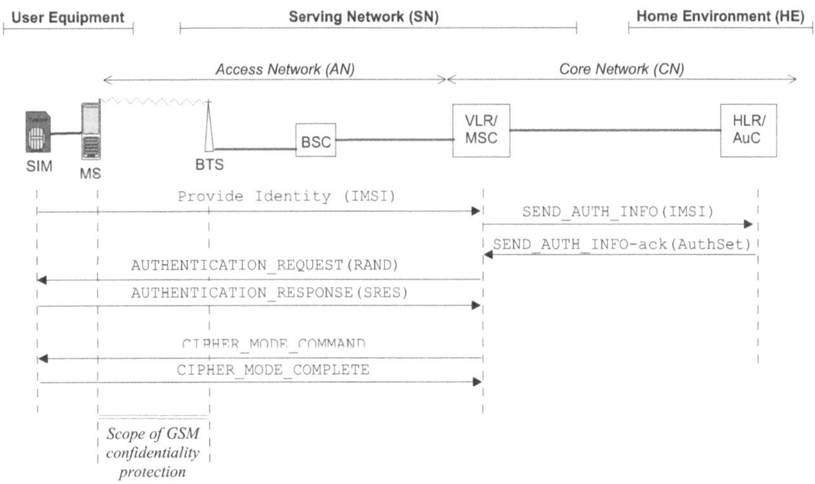

Figure 7.1 Outline of the GSM Access Security Architecture

7.1.2 System Architecture

The GSM AKA protocol is native to both the GSM and the GPRS systems. But, while GSM and GPRS share the AKA protocol, there are differences in the communications protection. This is partially due to the fact that GSM is a synchronous and symmetric (uplink/downlink) circuit-switched system, while GPRS provides packet-switched services. At the lowest layers, the radio channels used in GPRS are synchronized, but at the packet level the connection is asynchronous and directional (uplink/downlink).

In GSM, one provides data confidentiality protection for the radio link only (MS⇆BTS). In GPRS, the protection covers the whole range from the MS to the SGSN in the core network.

The following entities are involved in the GSM access security architecture:

- **GSM SIM**
  The GSM SIM contains the authentication secret K_I, the subscriber identity IMSI and the AKA algorithms A3 and A8. The GSM SIM will receive the challenge ($RAND$) and compute the Signed Response ($SRES$) and the session key K_C.

- **Mobile Equipment (ME) / Mobile Subscriber (MS)**
  The ME contains the A5 and GEA cipher algorithms and it will be the termination point for the communications encryption. The ME receives the session key (K_C) from the GSM SIM. All communication to/from the GSM SIM is via the ME. The ME with a SIM is called an MS.

- **Base Tranceiver Station (BTS)**
  The BTS is the termination point for communications security in GSM. It contains A5 cipher units. It will receive the cipher key K_C from the MSC (via the BSC).

- **The Serving GPRS Support Node (SGSN)**
  The core network node SGSN is the termination point for communications security in GPRS. A 2G SGSN contains GEA cipher units.

- **The VLR/MSC and the SGSN**
  The VLR/SGSN is the network-side termination for the GSM AKA challenge-response algorithm. It will receive the GSM AKA credentials, commonly called a "triplet', from the HLR/AuC.

- **HLR/AuC**
  The Authentication Centre (AuC) is a subsystem within the HLR and it is the entity where the per-subscriber long-term credentials are stored and where the per-subscriber AKA credentials are computed (A3/A8). The long-term credentials correspond to the credentials at the GSM SIM (IMSI and K_I). The HLR forwards AKA credentials (*triplet*) to the VLR/SGSN for roaming subscribers upon request.

The *triplet* is also known as an *authentication set*. The Authentication Set is defined as:

$$AuthenticationSet = \{RAND, SRES, K_C\}$$

The $RAND$ is a 128-bit wide random challenge, $SRES$ a 32-bit wide signed response and K_C is a 64-bit wide session cipher key.

7.1.3 The A3 and A8 Cryptographic Functions

The algorithms used in the GSM AKA protocol is called the A3/A8. The external interface to A3/A8 algorithms is found in Annex C in TS 43.020 [24].

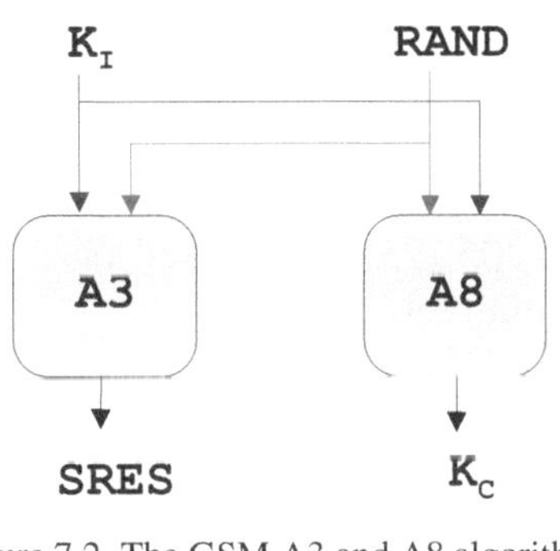

Figure 7.2 The GSM A3 and A8 algorithms

The purpose of Algorithm A3 is to allow authentication of a mobile subscriber's identity. The A3 function computes an expected response $SRES$ from a random challenge $RAND$ sent by the network. Algorithm A3 is controlled by the secret authentication key K_I. The A3 algorithm is physically located in the SIM and at the AuC. The A8 function takes exactly the same inputs as the A3 algorithm. The output from the A8 function is the secret session key K_C. As with the A3 algorithm, the A8 is physically located on the SIM and at the AuC. The A3 and A8 algorithms are not truly separate functions, and the implementations will tend to use one MAC function to compute both. Figure 7.2 shows the A3 and A8 functions. The A3/A8 implementation is not standardized at all and each operator may implement their own A3/A8.

7.1.4 The COMP128 Implementations

The A3 and A8 algorithms are considered to be operator specific. Nevertheless, the GSM Association has made available reference implementations of the A3/A8 algorithms. The "standard" implementations are collectively known as the COMP128 algorithms:

- **COMP128-1**: The "original" COMP128 algorithm. Designed to produce K_C with 54 significant bits. The algorithm was broken already in 1998 [49] and one is strongly advised against using it.

- **COMP128-2**: No serious flaw has been reported for the COMP128-2 algorithm. The algorithm produces a K_C with 54 significant bits, which is too short for real-life security.

- **COMP128-3**: This algorithm is identical to COMP128-2, but one has now allowed the key K_C to consist of 64 significant bits.

- **COMP128-4**: This algorithm is known as "GSM Milenage", and is specified in TS 55.205 [25]. It is based on functions from the 3GPP MILENAGE algorithm set.

Only COMP128-3 and GSM Milenage is really suitable for deployment today. However, given that one may run UMTS AKA over GERAN it would be *much* better to use a UICC/USIM and run the UMTS AKA protocol. The 3GPP has defined conversion functions for the key material for this case [10].

7.1.5 The A5 Stream cipher

The external interface to the A5 algorithm is given in TS 43.020 (Annex C.1 in [24]). The A5 algorithms must be common to all GSM networks and all mobile devices. Currently, there are four fully defined implementations of the A5 algorithm/interface (A5/1, A5/2, A5/3 and A5/4). Figure 7.3 depicts the algorithm interface and Figure 7.4 shows how it is used in the system.

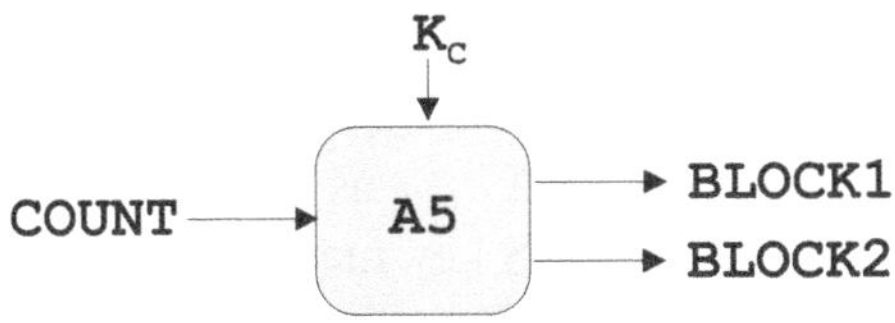

Figure 7.3 The GSM A5 Algorithm Interface

The $COUNT$ parameter is the TDMA frame number, which ensures explicit synchronization with the radio link. The key stream (BLOCK) data is XORed with the payload in the uplink and downlink *Normal Burst* frames as identified by COUNT. The A5 algorithm must be able to output a set of $BLOCK$ data pr burst; i.e. it must complete one cycle in less than 4.615 ms.

The coding of $COUNT$ also means that the output from A5 will repeat when the frame number is exhausted. The frame number covers $26 \times 51 \times 2048$ bursts, each lasting 4.615 ms. It thus takes approximately 209 minutes to exhaust the key stream. Repetition of the key stream is cryptographically highly undesirable. The only way to avoid repeating the pattern is to renew the key K_C, but this cannot be done while in connected mode in GSM.

The A5 algorithms:

- **A5/1** – The "standard" A5 implementation.
  The A5/1 algorithm is cryptographically broken. The first realistic attack was presented in "Real Time Cryptanalysis of A5/1 on a PC" [41] as far back as in 1999. Modern "rainbow table" attacks are quite efficient and A5/1 should no longer be the default algorithm.

- **A5/2** – This is the CoCom version of A5.
  It was designed to be sufficiently weak to allow export to the then east block countries. It is *completely broken*, and practical attacks exist [35]. The 3GPP and the GSM Association have officially deprecated the algorithm (The official network cut-off date was 2006.06.01).

- **A5/3** – This version is derived from the 3G KASUMI algorithm.
  The A5/3 algorithm is designed around the KASUMI block cipher primitive. The A5/3 mode-of-operation uses the 64-bit key K_C as input to the KASUMI core. The KASUMI core gets $K_C||K_C$ as the controlling key. No known practical attack exists against the A5/3 algorithm, but it is a 64-bit design and as such it is not future-proof.

- **A5/4** – This is the 128-bit key version of the A5/3 algorithm.
  The A5/4 is almost identical to the A5/3 function save for the fact that one uses a full 128-bit key Kc_{128} as the controlling key. The Kc_{128} key is derived from the UMTS key material (CK, IK) using the 3GPP standard KDF (see Appendix B in TS 33.102 [10]). The A5/4 function can only be used for the UMTS security context. Thus, the subscriber must have a UICC/USIM and one must run the UMTS AKA protocol.

In early 2013, the A5/1 was still the workhorse, but gradually we are seeing deployment of A5/3. The A5/4 is an ideal choice for subscribers with UICC/USIM, but A5/4 will probably not be available for another few years.

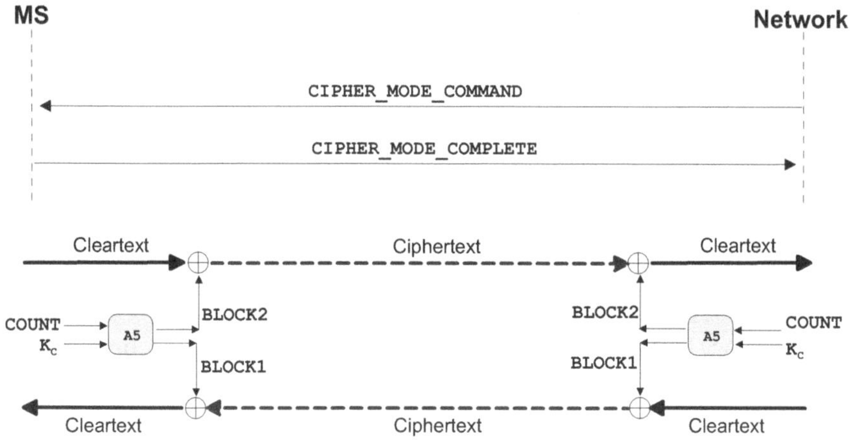

Figure 7.4 Ciphering setup and how the A5 algorithm is used

7.1.6 The GEA Cipher Functions

The GPRS Encryption Algorithm (GEA) is a familia of cipher functions used in 2G GPRS. The GEA algorithms are located in the ME and the SGSN. There exists four different GEA functions (GEA, GEA2, GEA3, GEA4).

The original GEA function was intentionally weakened by only using 54 of the available 64-bit key space. The GEA2 rectified this, and it is essentially the original GEA algorithm without the artificial weakening. The GEA3 algorithm is equivalent to A5/3 in that they both use the KASUMI core function with K_C taken twice as the input key. Likewise, the GEA4 algorithm is equivalent to the A5/4 algorithm, but this time with the K_{c128} key.

7.1.7 Outline of the GSM AKA Protocol

The GSM AKA protocol is, as is all the 3GPP AKA protocols, a two-stage protocol. The GSM AKA protocol is outlined in Figure 7.5 in Alice–Bob notation. The principal parties are MS, SN (VPLMN) and HE (HPLMN). The sequence starts with a triggering event, which in the example is an initial registration (the location updating signaling messages have otherwise been omitted). The registration event includes identity presentation with IMSI.

1. $MS \rightarrow SN$: LOC_UPDA_REQ($IMSI$)
2. $SN \rightarrow HE$: SEND_AUTH_INFO($IMSI$)
3. $HE \rightarrow SN$: SEND_AUTH_INFO(($RAND, SRES, Kc$))
4. $SN \rightarrow MS$: AUTH_REQ($RAND, CKSN$)
5. $MS \rightarrow SN$: AUTH_RESP($SRES$)

Figure 7.5 The GSM AKA protocol (abridged message names)

We shall not go into too much detail here, and we have already looked into the A3/A8 AKA algorithms. What matters to us is basically this:

- **The AKA protocol is network initiated and optional**
  The GSM AKA is always network initiated (VPLMN). Furthermore, it is optional for the network to carry out the AKA protocol.

- **Unprotected triplet request and forwarding**
  There is no standardized cryptographic confidentiality and integrity protection mechanism for the channel between HE and SN in the GSM environment. It is therefore likely that the request (SEND_AUTH_INFO) is not authenticated. Likewise, the response with the session credentials (triplet) will likely be forwarded without any protection.

- **Two-stage protocol with no real home control**
  The GSM AKA is a two-stage protocol. The HPLMN may be offline during the challenge-response. There is an option to report AKA failure, but it is not mandatory for use. This means that the HPLMN has limited control over the subscriber while roaming onto a foreign network.

- **No state and no freshness guarantee (no replay protection)**
  There is no state information in the GSM AKA protocol to distinguish different protocol runs. That is, there is no record kept about which challenges have been used and there is no way to ascertain freshness

either. This means that if an intruder observes the $(RAND, SRES)$ exchange, which after all is in plaintext, then he/she may replay that sequence later.

- **No network authentication**
  The GSM AKA protocol only provides unilateral authentication. The subscribers have no assurance of either the VPLMN or the HPLMN. This is a major shortcoming.

- **Inadequate response length and severe key problems**
  The signed response $(SRES)$ is only 32-bit long. This is too short and it should have been at least 64-bit long. The derived key material, the 64-bit K_C, is also insufficient by todays standards. The derived key material should have been at least 128-bit long. The key scope in GSM/GPRS is only for ciphering, integrity protection is completely missing. Key "validity" is also an issue. A key should never be used for more that one algorithm, but there is no restriction on key re-use at all in GSM/GPRS. This is bad since the K_C may potentially be used for all of A5/1, A5/2, A5/3, GEA, GEA2 and GEA3. We have also seen that the $COUNT$ parameter will repeat after approx. 3.5 hours, but there is no explicit restriction on the lifetime of the K_C key.

- **Re-keying requires AKA re-run**
  In GSM, there is no local re-keying procedure. This means that one must either move the key around with the user (to the handling BTS) or re-authenticate every time the MS moves to a new cell.

7.1.8 Summary

The GSM AKA algorithm and the GSM Security Context is flawed and inadequate. The security is rather weak and and the subscriber privacy depends on a scheme (using TMSI) that is totally inadequate in the presence of an active intruder.

7.2 Outline of the UMTS Access Security Architecture

7.2.1 Background

The UMTS system architecture encompasses both packet-switched and circuit-switched services in the same system. The GSM and GPRS core network nodes have all been retained, but as of Release 5 and onwards the HLR/AuC has been replaced by the HSS, which is a superset of the HLR/AuC. The GSM SIM has been replaced by the UICC smart card and the USIM subscriber module. The UMTS radio access network, called UTRAN, consists of Radio Network Control (RNC) nodes and base stations, which are called NodeB (NB). A UMTS core network can, and very likely will, support GERAN access networks.

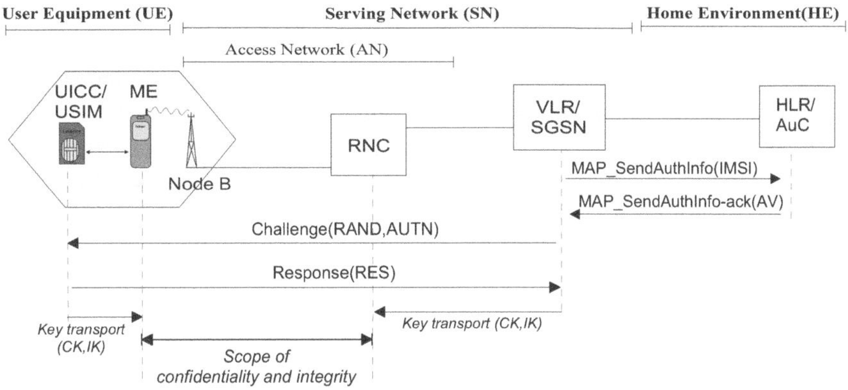

Figure 7.6 The UMTS Access Security Architecture

7.2.2 System Architecture

One notable difference from the GSM/GPRS system is that access security protection termination point now has been unified. In 2G, the CS part was terminated in the basestation (BTS) while the PS part was terminated in the SGSN. In UMTS, one realized that termination at the basestation, called NodeB in UMTS, would be insufficient. At the same time, the GPRS approach with termination in the core network proved problematic too. The problem is that the low level radio processing needs access to some parts of the plaintext. Thus, in GPRS one had to handle low-level radio access information in the core network. The main radio access decision point in UMTS is the Radio Network Controller (RNC). The RNC plays much the same role as does the BSC in 2G, but the RNC is a more autonomous network element. The RNC also handles both CS and PS connections and it is the access security termination point.

The UMTS access security architecture is primarily captured in TS 33.102 [10]. Other standards that play a role in the UMTS access security architecture is TS 21.133 [2], which has a threat and risk analysis of the environment facing the 3G network. Figure 7.6 presents an overview of the UMTS access security architecture. The UMTS security architecture is also treated in depth in [167, 196].

7.2.3 Network Security

To protect the IP-based protocols in the core network one decided to use native IP security mechanisms and profile them for use within the 3GPP core network. The natural solution was to use IPsec (then RFC 2401). The 3GPP way of applying IPsec in the UMTS core network is called NDS/IP (TS 33.210 [13]). Figure 7.7 depicts the NDS/IP architecture.

The NDS/IP solution represents a proper subset of IPsec, but with some strengthening of requirements to guarantee security and to provide better and safer inter-operability. The NDS/IP architecture is primarily designed to protect inter-operator communication, with use of security gateways (SEGs) at the network borders. NDS/IP can also be used within an operator network, and this use is increasing in LTE. The NDS/IP security services are:

- Data Integrity and Data origin authentication
- Message replay protection
- Data Confidentiality (optional)
- (Limited) protection against traffic flow analysis

NDS/IP has standardized on the following in IPsec:

- **Use of tunnel-mode**
  Tunnel-mode is required for inter-operator Za-interface. Transport-mode *may* be used for the intra-operator Zb-interface.
- **Mandatory use of Encapsulated Security Payload (ESP)**
  IPsec has two protection protocols: The ESP protocol and the Authentication Header (AH) protocol. NDS/IP requires both data integrity- and data confidentiality protection. Only the ESP protocol can provide this.
- **Mandatory support for integrity protection**
  The ESP protocol allows for selection of NULL integrity and NULL confidentiality simultaneously. The 3GPP has mandated use of integrity protection for all NDS/IP protected connections. Furthermore, replay protection is made mandatory.
- **Confidentiality protection**
  Data confidentiality protection is not mandatory in NDS/IP, but when it is used it shall always be used in conjunction with integrity protection.

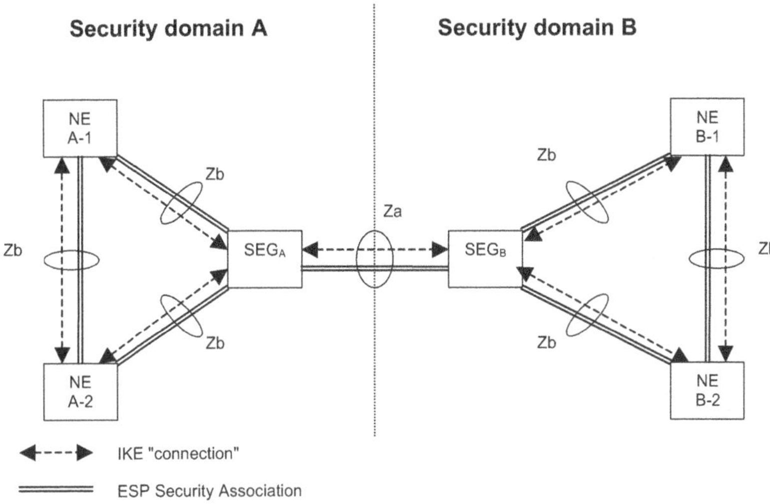

Figure 7.7 The NDS/IP architecture

IPsec permits some weak cryptographic algorithms to be used, but these are explicitly not allowed in NDS/IP. The basic NDS/IP architecture relies on the use of pair-wise pre-shared secrets as the basic long-term security credential. To avoid scalability problems, a PKI-based extension has been added to the basic NDS/IP architecture. This extension to the NDS/IP architecture is specified in TS 33.310 [14]. The PKI extension can coexist with the basic pre-shared secret scheme.

7.2.4 The Authentication Vector

The cryptographic basis for subscriber authentication is a 128 bit wide pre-shared secret key, K, that resides on/in the USIM and at the AuC. The key K and the permanent subscriber identity $IMSI$ is the security credential basis for the UMTS AKA protocol. Each UMTS AKA protocol run is associated with a security credential set called the Authentication Vector (AV). The AV is subscriber specific and directly associated with the (K,$IMSI$)-tuple.

$$AUTN := (SQN \oplus AK)||AMF||MAC{-}A$$

$$AV := (RAND, XRES, CK, IK, AUTN)$$

The AV corresponds roughly to the GSM triplet, but contains more elements to facilitate mutual authentication. The use of sequence numbers in the $AUTN$ element may allow an intruder to track the subscriber. To preserve anonymity, the SQN value may therefore be concealed by XORing it with an anonymity key AK. The AK is generated by the $f5$ function.

7.2.5 The UMTS Cipher Primitives and the f8/f9 Functions

7.2.5.1 The UMTS Cipher Primitives

TS 33.105 specifies a data confidentiality function ($f8$) and a data integrity ($f9$) function. The $f8$ and $f9$ interfaces must be fully standardized since the functions are placed in the ME and in the RNC. To allow multiple alternative algorithms one has defined the UEA and UIA algorithm identifiers. The UEA and UIA identifiers are 4 bits wide and thus allow for encoding of up to 16 different algorithms for both data confidentiality and data integrity. This includes NULL functions, although the use of a NULL functions for $f9$ (integrity) is not permitted.

In UMTS, one started off with one cipher primitive for data confidentiality and data integrity protection. This primitive is called KASUMI (*mist* in Japanese) and it is based on the MISTY1 cipher from Mitsubishi. The KASUMI primitive is based on using a 128-bit key, but is internally limited to a 64-bit block length [17]. This is a weakness and this is also why KASUMI is not on the list of approved cipher primitives for LTE. Some papers have claimed to have "broken" KASUMI [86]. The attacks are significant, but KASUMI cannot be said to be broken in practical use in the intended setting. However, the problems associated with 64-bit internal nature was recognized by the standards committee (3GPP SA3) as not future proof and so a second cipher was commissioned. This cipher is called SNOW-3G [22] and it is based on the SNOW 2.0. The use of $f8/f9$ as interface functions continued with SNOW-3G. The SNOW cipher family is not, like KASUMI, a block cipher but rather a native stream cipher. Furthermore, SNOW-3G is also a native 128-bit design and this ensured that SNOW-3G is one of the LTE approved cipher primitives. The books [173, 196], contains a much more detailed account of the UMTS security architecture and is recommended reading for the interested reader.

7.2.5.2 The UMTS Confidentiality ($f8$) Function

The data confidentiality function $f8$ takes a bearer identifier ($BEARER$), a direction indication ($DIRECTION$), a time dependent input ($COUNT$-

C), and a length parameter ($LENGTH$) as input. The $f8$ function operates under control of a 128-bit key CK. The output of $f8$ is a pseudo-random keystream. Figure 7.8 illustrates the use of $f8$ to encrypt plaintext by applying a keystream using a bitwise XOR operation. The plaintext is recovered by generating the same keystream using the same input parameters and applying it to the ciphertext using a bitwise XOR operation. A thorough discussion of the $f8$ is found in Chapter 6 of [196].

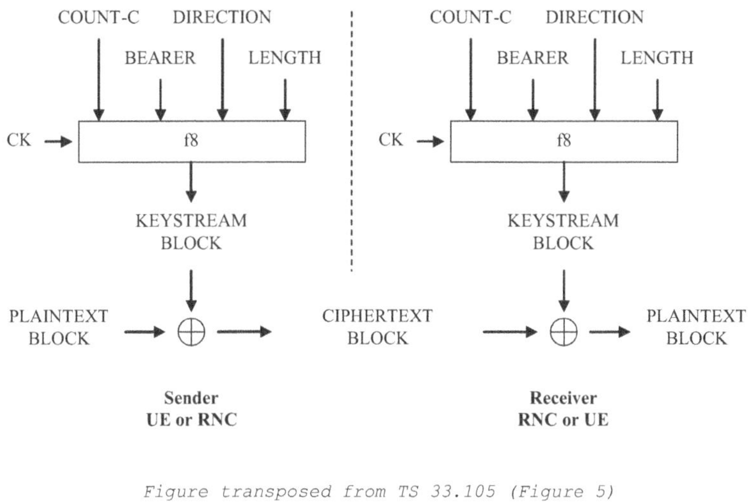

Figure 7.8 The $f8$ (confidentiality) function

7.2.5.3 The UMTS Integrity ($f9$) Function

The input parameters to the $f9$ function are the integrity key (IK), a time dependent input ($COUNT\text{-}I$), a random value generated by the network side ($FRESH$), the direction bit ($DIRECTION$) and the signaling data ($MESSAGE$). The output is the message authentication code for data integrity ($MAC\text{-}I$) which is appended to the message when sent over the radio access link. The receiver computes $XMAC\text{-}I$ on the messages and compares the integrity check values to verify that the message has not been altered. Figure 7.9 illustrates the use of the $f9$ function. Integrity protection in UMTS is limited to the coverage of system signaling messages between the MS and the RNC. The fact that user data is not integrity protected represents a problem under certain circumstances. In particular, there will be situations where encryption is not available and where the integrity protection is the only line of defense. So the lack of user data integrity protection introduces a problem in countries where encryption is not allowed or otherwise not available, since it opens up the possibility for an attacker to illicitly modify user data or effectuate session hijacking.

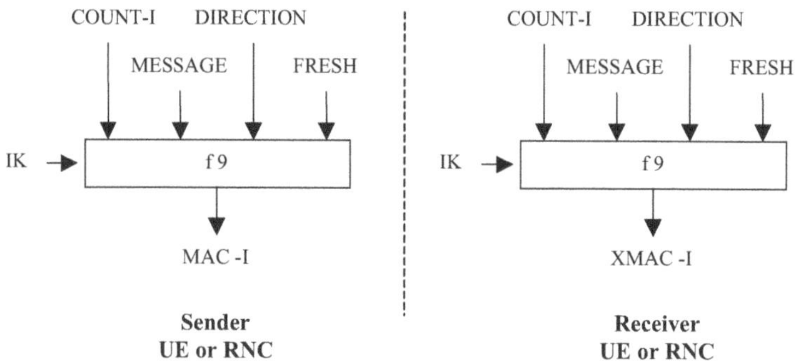

Figure transposed from TS 33.105 (Figure 6)

Figure 7.9 The $f9$ (integrity) function

The $MAC\text{-}I$ value used is only 32-bit wide. This would under most circumstances be considered insufficient. However, for the generally very short signaling messages even a 32-bit MAC increases the signaling load substantially. Still, it would have been safer and better to use at least a 64-bit checksum value. A thorough discussion of the $f9$ function is found in Chapter 6 of [196].

7.2.6 The UMTS AKA Functions

7.2.6.1 Overview

The UMTS AKA functions and parameters are defined in TS 33.105 "Cryptographic algorithm requirements" [11]. The specification only defines the input/output and functional requirements of the functions. The pseudo-random number generating function ($f0$) is only present at the AuC. The cryptographic functions ($f0$–$f5*$) used in the AKA procedure are located and used in the USIM and the AuC. The UMTS operators are free to choose any algorithm they want provided it complies with requirements in TS 33.105 [11]. The 3GPP has also developed an example set of functions called the MILENAGE algorithm set [18–21, 23]. The MILENAGE algorithms is, in practice, the standard algorithm set for the AKA function. MILENAGE is built around the Rijndael block cipher (Rijndael was later adopted by NIST as the AES [197]). Table 7.1 depicts the UMTS AKA functions and their use. A set of AKA conversion functions (for backwards compatibility with GSM) is defined in TS 33.102 [10].

Table 7.1 UMTS AKA functions

Function:	Purpose/Usage:
$f0$	Random challenge generating function
$f1$	Network authentication function
$f1*$	Re-synchronization message authentication function
$f2$	User authentication function
$f3$	Cipher key derivation function
$f4$	Integrity key derivation function
$f5$	Anonymity key derivation function for normal operation
$f5*$	Anonymity key derivation function for re-synchronization

7.2.6.2 The Random Challenge Generating Function ($f0$)

The $f0$ function is not standardized and is only present at the AuC.

$$f0(internal\_state) \rightarrow RAND$$

7.2.6.3 The Network Authentication Function ($f1$)

The network authentication function should be a MAC function [11]. It must be computationally infeasible to derive K from knowledge of $RAND$, SQN, AMF and $MAC\text{-}A$. Obviously, it must also be infeasible to derive $MAC\text{-}A$ from knowledge of $RAND$, SQN and AMF alone.

$$f1_K(SQN, RAND, AMF) \rightarrow MAC\text{-}A$$

The $f1$ function is used to verify the authenticity of the challenge. The USIM, upon receiving the challenge, can verify that the challenge data originated with an entity which possess K. The USIM must also assure itself that the challenge is still valid (as in fresh/recent). This is verified with the sequence number (SQN) included in the challenge. The USIM maintains state information of the present and past SQN values. The exact procedure of SQN verification has not been standardized. A set sequence number management schemes are proposed in TS 33.102 Annex C [10], but these schemes are neither mandatory for implementation nor for use. If the challenge has expired, the USIM will initiate the re-synchronization procedure.

We will not discuss the re-synchronization here. The interested reader should consult [10, 11] for more information on the the re-synchronization procedure.

7.2.6.4 User Authentication Function ($f2$)

The user authentication function should be a MAC function [11]. The output from $f2$ is the RESponse value RES. The term $XRES$ is used to denote the expected response, but it is actually the same identifier as RES. It must be computationally infeasible to derive K from knowledge of $RAND$ and RES. It must also be infeasible to derive RES from knowledge of $RAND$ alone. The requirements must hold even when a large set of ($RAND$,RES) pairs have been observed by an intruder.

$$f2_K(RAND) \rightarrow RES$$

The VLR/SGSN verifies that $RES = AV.XRES$. If verification succeeds, the VLR/SGSN considers the USIM to be authenticated.

7.2.6.5 Key Derivation Functions ($f3$ and $f4$)

It must be computationally infeasible to derive K from knowledge of $RAND$ and the derived keys. It must also be infeasible to derive the keys from knowledge of $RAND$ alone. The USIM computes CK and IK during the authentication sequence.

$$f3_K(RAND) \rightarrow CK$$

$$f4_K(RAND) \rightarrow IK$$

7.2.6.6 Anonymity Key Derivation Function ($f5$) (for normal operation)

It must be computationally infeasible to derive K from knowledge of $RAND$ and AK. It must also be infeasible to derive AK from knowledge of $RAND$ alone. The USIM must derive AK *before* it can verify the $AUTN.SQN$ value.

$$f5_K(RAND) \rightarrow AK$$

There is another $f5$ function defined for the re-synchronization case, but we will not discuss this function here (see instead [10, 11]).

7.2.7 Outline of the UMTS AKA Protocol

7.2.7.1 The UMTS AKA Entities

There are three principal entities in the UMTS AKA sequence.

- **USIM**
  The USIM is part of the User Equipment (UE) and it is the USIM that responds to the challenge. The USIM will forward the session keys (CK, IK) to the ME subsequent to UMTS AKA execution.
- **VPLMN**
  The VPLMN is a collection of physical nodes and services. It is also called the Serving Network (SN). The SN node active in the UMTS AKA execution is the VLR/SGSN.
- **HPLMN**
  The HPLMN is a collection of physical nodes and databases. It is also called the Home Environment (HE). The HE node active in the UMTS AKA execution is the HLR/AuC (HSS).

7.2.7.2 The UMTS AKA Protocol; The AV Forwarding

The forwarding of the AV from the HLR/AuC to the VLR/SGSN is essentially similar in GSM and UMTS. Figure 7.6 shows this. The MAP protocol [7] is used for the transport of the security credentials and it can handle both the GSM triplets and the UMTS AV. A major drawback to using the SS7-based MAP protocol is that there are no security mechanisms available. There is no security protocol a la IPsec or TLS for the SS7 protocol stack and the MAP protocol does not itself provide any security services. It is possible to transport the SS7 messages over IP (SIGTRAN) and one may then use IPsec to protect the signaling, but otherwise the security and privacy sensitive MAP protocol messages will be forwarded in plaintext. This is an obvious and rather devastating weakness in the UMTS security architecture.

7.2.7.3 The UMTS AKA Protocol; The Challenge-Response

The UMTS challenge-response procedure is similar to the GSM challenge-response procedure, and the message flow is more-or-less identical. Figure 7.6 depicts the UMTS AKA scheme.

The VLR/SGSN initiates the local AKA procedure by sending the challenge message (`AUTHENTICATION_REQUEST`) that contains the random challenge $RAND$ and the authentication token $AUTN$. The UE verifies that the $AUTN.MAC\text{-}A$ is correct for the $(RAND, AUTN)$-challenge and that the challenge is fresh (i.e. not used before or not otherwise expired). For the successful case, the UE will respond with the AUTHENTICATION_RESPONSE message. The VLR/SGSN then verifies that the response (RES) matches the expected response ($AV.XRES$). The successful procedure is executed in a single round trip (one-pass) and the message exchange is identical to the corresponding GSM scheme.

The UMTS AKA protocol is outlined below in the common Alice–Bob notation. The parties are UE, SN and HE. The sequence starts with a triggering event, which in the example is the location updating (LU) procedure. The LU procedure includes identity presentation. The message names have been abridged.

1. $UE \rightarrow SN$: LOC_UPDA_REQ($IMSI$)
2. $SN \rightarrow HE$: SEND_AUTH_INFO($IMSI$)
3. $HE \rightarrow SN$: SEND_AUTH_INFO(AV)
4. $SN \rightarrow UE$: AUTH_REQ($RAND, AUTN$)
5. $UE \rightarrow SN$: AUTH_RESP(RES)

Figure 7.10 The UMTS AKA protocol

There are well-defined procedures for the failure cases ($MAC-A$ failure) and for re-synchronization (invalid SQN), but we will not discuss those here (see instead [10, 196]).

7.2.8 Weaknesses, Shortcomings and Omissions

UMTS access security inherited a lot from GSM/GPRS, and sometimes the history shows all too well. Some of the more obvious problems are listed below:

- **Authentication**
  - *Delegated authentication*
    The HPLMN delegates **all** authentication authority to the VPLMN. This is a serious shortcoming of the UMTS security model unless the required trust can be justified.
  - *Unauthenticated plaintext transfer of security credentials*
    By default there is no realistic way to secure the MAP protocol.
  - *Sequence Number Management*
    The sequence number management scheme in UMTS is not standardized. There is a real risk that the SQN scheme will not be able to provide replay-protection.

  - *Continued use of GSM SIM in UMTS networks*
    Use of GSM SIM lowers the security level down to the GSM level.

- **Key Distribution**
  - *Unprotected (CK, IK) transport from the VLR/SGSN to the RNC*
    There is no default protection.
  - *Unprotected (CK, IK) transport from the USIM to the ME*
    There is no security on the USIM – ME interface.
- **Rouge Shell and NodeB Compromise**
  - *Compromised ME (Rouge Shell)*
    The new smart phones are advanced computer platforms. They are vulnerable to attacks and may be compromised. There is no standardized protection in UMTS against a compromised ME.
  - *Compromised NodeB*
    The NodeBs are widely distributed and will often be located in hostile environments, yet there is no standardized system protection for the NodeBs.
- **Link Layer Protection**
  - *Limited integrity protection on the link layer*
    The 32-bit integrity check value is too short for comfort.
  - *Protection Coverage*
    Since confidentiality protection is an option, it is problematic that there is no integrity protection for user data.
- **Subscriber Privacy**
  - *Permanent subscriber identity routinely exposed*
    The $IMSI$ is routinely exposed over the Uu-interface. There is no means of preventing this in UMTS.
  - *Permanent subscriber identity/location not protected*
    An active attacker may page with $IMSI$ and the UE is obliged to answer the paging since there is no way for the UE to determine if the request is valid.
  - *Temporary identity may be correlated with permanent identity*
    There is no formal requirement in UMTS on how to generate the $TMSI$. The $TMSI$ may be structured and/or allocated sequentially (local to the VLR/SGSN).

- *No requirement on expiry of the temporary identity*
  The $TMSI$ may be used for a prolonged period. It may be possible to track an anonymous user for long periods.
- *Sequence Number Management*
  There is a risk that the SQN may leak information to an intruder. Use of AK and $f5$ should prevent this.

7.2.9 Summary

The UMTS security architecture represents true improvement over the GSM security. The cryptographic shortcomings have been addressed, but architectural constraints and backwards compatibility requirements have impeded UMTS security.

This is problematic since the environment (threats and risks) facing UMTS is much more complex than the simple speech-oriented environment that the GSM system existed in. Still, native UMTS security is reasonable if all options are exercised.

With respect to subscriber privacy, we unfortunately have that case that while data privacy over-the-air has been improved, the problems with identity privacy and location privacy etc. essentially remains the same as in GSM/GPRS.

7.3 Outline of the EPS Security Architecture

7.3.1 Background

The Long Term Evolution (LTE) system consists of a radio system (E-UTRA) and an Evolved Packet Core (EPC) network system. The LTE architecture is also an All-IP Network (AIPN). The Signaling System No.7 (SS7) protocols previously used in the core network in GSM/GPRS and UMTS are now all gone. An important aspect is that LTE is **not** backwards compatible with the GSM SIM smart card or the GSM AKA protocol. To get access to LTE, one *must* be able to run the UMTS AKA protocol.

The UMTS AKA protocol used in LTE is a component in the so-called EPS AKA protocol. This allows the subscribers to use standard UICC/USIM in the LTE architecture. The UICC/USIM is the only subscriber identity module available for LTE. The EPS acronym stands for Evolved Packet System and is used in the 3GPP specifications for denoting a 4G system.

The new LTE radio system, E-UTRA, is providing true mobile broadband services. The data rates can easily reach 100 Mbps, but E-UTRA is still not fully compliant with the IMT-Advanced vision of a 4G system. An enhanced version of LTE, called LTE-Advanced, brings the radio access network up in full compliance with the IMT-Advanced vision of 4G.

7.3.2 System Architecture

The LTE system model is described in TS 23.401 [6] and to some extent in TS 23.002 "Network Architecture" [3]. These documents describe LTE and how LTE integrates with UMTS and GSM/GPRS. There are two main LTE scenarios; the roaming and the non-roaming case. The roaming case includes the home network (HPLMN) and the visited network (VPLMN) as separate administrative and operative domains. The non-roaming case is the case where both the HPLMN and the VPLMN coincide as depicted in Figure 7.11. The figure is transposed from TS 23.401, Figure 4.2.1-1.

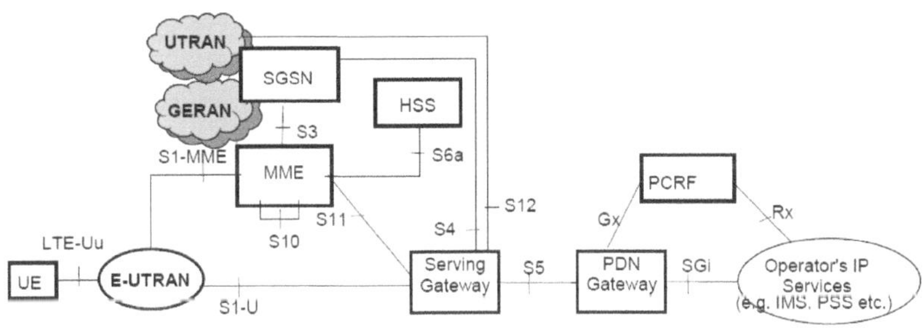

Figure 7.11 LTE System Architecture – Non-roaming case

The reference model also distinguishes between cases with local breakout (of the user plane traffic) and for cases where the user plane is always tunneled back to the HPLMN. Whether or not local breakout is permitted is an operator choice. In 2G GPRS and UMTS it has been the norm to tunnel all of the user plane traffic back to the home network. We shall mainly look at the roaming case since that provides the most generic view. Figure 7.12 depicts the LTE roaming architecture w/user plane routed to the HPLMN. The figure is transposed from TS 23.401, Figure 4.2.2-1.

7.3.3 Network Security

In LTE, all network communication is packet-switched and over IP. The difficulties of providing security solutions for the SS7-based protocols can

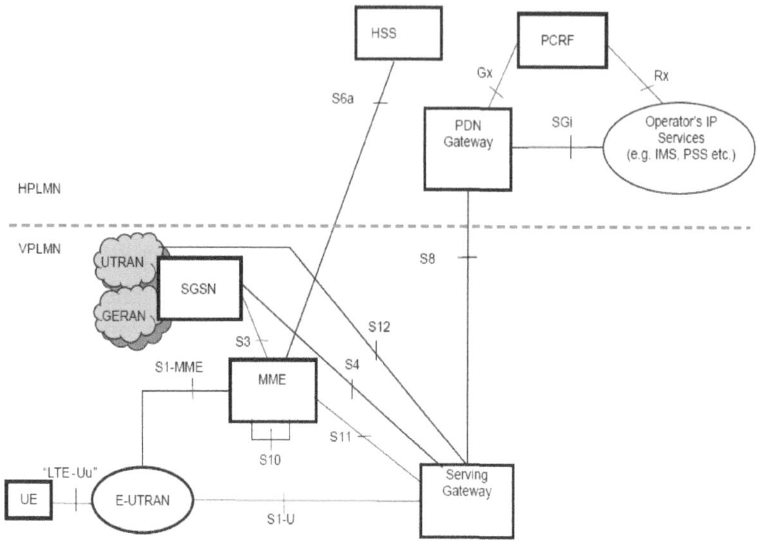

Figure 7.12 LTE System Architecture – Roaming case

now finally be overcome. The preferred solution is to use NDS/IP (see Chapter 7.2.2). NDS/IP is now used for both inter-operator (tunnel-mode) and intra-operator (tunnel-mode or transport-mode) traffic. The most important protocols to protect are the GPRS Tunneling Protocol (GTP) and the MAP replacement protocol. The MAP replacement protocol is functionally very similar to the SS7-based MAP protocol and is based on the AAA/DIAMETER framework.

7.3.4 The Basic Security Architecture

The LTE access security architecture is different from the UMTS access security architecture, but there are many similarities too:

- The USIM is retained unchanged.
- The HSS subscriber security handling is very similar.
- The entity authentication is very similar.
- The Authentication Vector (AV) concept is kept.

7.3.4.1 Stratums

The EPS security architecture differentiates between control plane and user plane traffic. In particular, all non-access related control plane traffic is handled separately in the "Non-Access Stratum (NAS)". The NAS has its own control plane protocol called the NAS protocol. Mobility management and security handling (EPS-AKA) in the E-UTRAN part of the VPLMN is handled by the NAS protocol.

The "Access Stratum (AS)" is all user plane associated traffic for the over-the-air interface (LTE-Uu). The access stratum reaches only between the ME and the basestation (eNB). This also means that the eNB will be the termination point for the AS security, and the eNB is an important part of the EPS security architecture.

The AS and NAS have separate security contexts with separate session keys. Both the `AS Security Context` and the `NAS Security Context` belong to the `EPS Security Context`.

7.3.4.2 The EPS Authentication Vector

The EPS Authentication Vector (EPS-AV) differs from the UMTS AV in that the UMTS session keys (CK,IK) is replaced by the 256-bit wide K_{ASME} key. The full EPS-AV is then:

$$EPS{-}AV = \{RAND, XRES, K_{ASME}, AUTN\}$$

The K_{ASME} is derived from the (CK,IK) keys by means of a standardized Key Derivation Function (KDF). The K_{ASME} is the basis for all keys in the EPS key hierarchy. Security contexts derived from EPS-AV with the EPS-AKA protocol are called *Native EPS security contexts*. It is possible to convert between LTE and UMTS security contexts. In LTE parlance, a UMTS security context is called a *Legacy security context*. A converted security context is called a *Mapped security context*.

No mapping between GSM security context and LTE security context is permitted. Thus, a security context created by the GSM AKA protocol *cannot* be used in LTE. This is also true for the case where a GSM security context is mapped to a UMTS security context; this context cannot later be mapped onto a LTE context.

7.3.4.3 The USIM and the Mobile Equipment (ME)

The USIM used in LTE does not know about LTE/EPS and consequently all security that goes beyond UMTS cannot be handled by the USIM. This means that the ME has got new responsibilities in LTE with respect to security. Amongst others the ME must be able to derive the master key K_{ASME} and the associated key hierarchy. The local session keys for the over-the-air protection (`AS Security context`) is also handled by the ME. The ME must also be able to verify that the EPS-AKA challenge has the "separation" bit set in the $AUTN$ part of the challenge.

7.3.4.4 The eNodeB (eNB)

The basestation in LTE is called eNodeB (eNB). The E-UTRAN network consists of mesh-connected eNBs. The eNBs are also connected to a Mobility Management Entity (MME) and a Serving Gateway (SGW). The MME and SGW replaces the SGSN in LTE. The MME handles control plane traffic and the SGW handles user plane traffic.

The eNB plays an active part in the access security. The eNB will store key material, it will encrypt and decrypt data and it will also derive key material. So the eNB must be able to process data securely and it must be able to maintain its own physical integrity. Section 5.3 in TS 33.401 [16] gives an overview over the security requirements for the eNB.

7.3.4.5 The Serving Gateway and the PDN Gateway

The two gateways used in LTE are the Serving Gateway (SGW) and the PDN Gateway (PGW). The gateways may be implemented in one physical node or in separated physical nodes. The user plane traffic is handled by the gateways. The PGW has the SGi-interface, which is the entry/exit point for all IP traffic to/from an LTE network.

7.3.4.6 The Access Security Management Entity (ASME)

The MME/ASME serves roughly the same purpose as the VLR/MSC and/or the SGSN. The Mobility Management Entity (MME) is a logical entity. Associated with the MME is the Access Security Management Entity (ASME). The MME/ASME are both hosted in a physical server. The ASME is the logical contact point for the EPS-AV requests over the S6a interface towards the HSS. The ASME also handles the EPS-AKA challenge-response procedure towards the UE (USIM). It is customary to refer to only the MME, even when the actual entity would be the ASME.

7.3.5 The EPS Security Contexts and Key Hierarchy

7.3.5.1 The Base Credentials

The basis is still the $(K,IMSI)$-tuple used in UMTS, and it is stored in the HSS and the USIM. The EPS-AV, which is based on the $(K,IMSI)$-tuple, is the basis for the `EPS Security Context`. The EPS-AV is created at the HSS and forwarded to the handling MME upon request. The forwarding takes place over the S6a-interface.

The K_{ASME} is new to the EPS-AV. In contrast to the (CK, IK) key pair that it replaces, the K_{ASME} is itself never used as a session key. Instead it is used as a key-deriving-key, forming the basis for the EPS key hierarchy. At the subscriber side, the ME must derive both the K_{ASME} and the keys created from the K_{ASME}, while at the network side the HSS derives K_{ASME} and the MME derived the various other keys in the EPS key hierarchy. Another difference in the EPS-AV is that the AMF part of the $AUTN$ now has a reserved "separation" bit. An ME accessing E-UTRAN must verify that the AMF "separation" bit is set to **1**.

7.3.5.2 The Security Contexts

There are three native security context types in EPS. First, we have the `EPS Security Context` that is established directly through running the EPS-AKA protocol. The `EPS Security Context` is based on the subscriber identity (IMSI/GUTI), the VPLMN identity and the master key K_{ASME}. It includes the `NAS Security Context` and (on-demand) the `AS Security Context`.

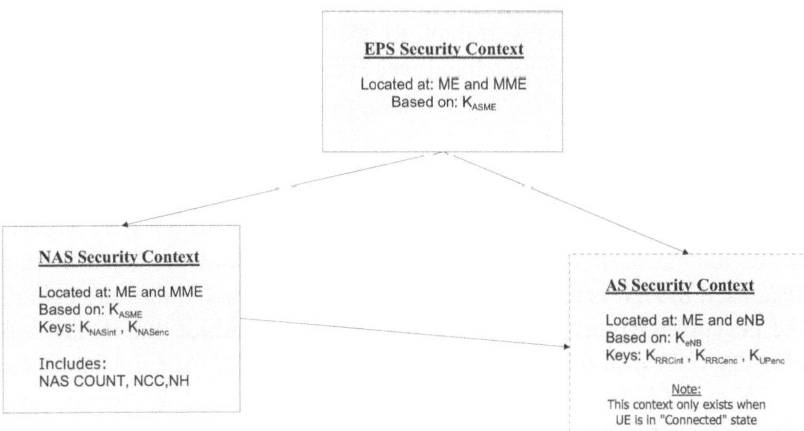

Figure 7.13 EPS Context Hierarchy

Figure 7.13 shows the relationship between the contexts. The `AS Security Context` only exists when the UE is "connected" to the network while the `NAS Security Context` also exists when the UE is "idle".

The `EPS Security Context` may exist also when the UE is not registered. A key set identifier ($eKSI$) is then used as a reference to the existing `EPS Security Context`. The `EPS Security Context` may even exist when the ME is physically turned off. During power-on, the ME will retrieve the stored context and verify that the originating UICC/USIM is still present. If that is not the case, the ME will delete the `EPS Security Context`. Under no circumstance will the ME allow the K_{ASME} to leave the secure storage of the ME. A *native* (non-mapped) `NAS Security Context` may also be stored during power-off.

7.3.5.3 The EPS Key Hierarchy

In LTE, one needs independent key sets for the user plane (UP), the RRC protocol and the NAS protocol. The K_{ASME} master key key is 256-bit wide and all the other keys are derived from the K_{ASME}. All derived keys are 256-bit wide, but the session keys are truncated to 128 bit. Figure 7.14 outlines the key hierarchy.

All the key derivations are using the same standard KDF, but with different parameters and controlling keys. The key derivation is not itself difficult, but the rules for re-keying in the `AS Security Context` are indeed fairly complicated. We shall refer to TS 33.401 [16] for more information on these procedures. There are also other sources [134, 173] that explain the procedures. For our purpose, suffice to say that the entire `AS Security Context` is renewed for all handovers. In fact, to invalidate an existing `AS Security Context` the system triggers a handover event.

7.3.5.4 Algorithms

Given that the USIM is retained it is no surprise that the AKA algorithms used in LTE are the UMTS AKA algorithms. There is also a KDF algorithm (see Section 7.3.6.3). The data integrity and data confidentiality function interface used in LTE is the same as for UMTS ($f8/f9$), but they are called the EPS Encryption Algorithm (EEA) and EPS Integrity Algorithm (EIA) respectively. The only difference between $f8/f9$ and the EEA/EIA interfaces is that CK and IK has been replaced with a generic KEY parameter, where KEY is one of the session keys belonging to the `AS Security Context` and the `NAS Security Context`.

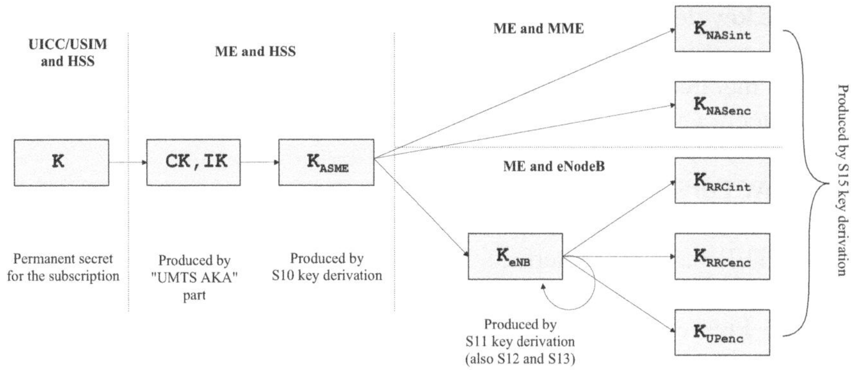

Figure 7.14 EPS Key Hierarchy

For data confidentiality, one has two basic cryptographic primitives, the familiar SNOW-3G [22] and AES in counter mode (CTR) [198]. There is also a Chinese algorithm called ZUC (see `http://zuc.dacas.cn/`). The three algorithms (SNOW-3G, AES and ZUC) are also used for the data integrity algorithm. Annexes B and C in TS 33.401 [16] contain more information on the algorithms and their use in LTE.

- EEA1/EIA1 is realized by SNOW-3G as the core primitive.
- EEA2/EIA2 is realized by AES as the core primitive.
- EEA3/EIA3 is realized by ZUC as the core primitive.

Formally, each of the EEA/EIA interfaces is prefixed with the key length (128-bit). Thus, the complete name for SNOW-3G used for data confidentiality is really 128-EEA1.

7.3.6 Outline of the EPS-AKA Protocol

There are three parts to the EPS AKA protocol, two of which is inherited from the UMTS AKA protocol.

- **The forwarding stage**
  The EPS-AV forwarding is reminiscent of the forwarding stage in GSM AKA and UMTS AKA. However, with LTE the transport protocol has changed from MAP to a AAA/DIAMETER-based protocol.
- **The authentication part** The basic authentication is very similar to the UMTS AKA authentication, and at the USIM side it is identical. There are some additions, like (weak) verification of the VPLMN identify and confirmation that the challenge belongs to an EPS-AV.

- **The key derivation and context setup part**
  This part is significantly different from its predecessors. There is a full key hierarchy and local re-keying no longer requires running the AKA protocol.

7.3.6.1 AV Forwarding

The AV forwarding takes place between the MME and the HSS over the DIAMETER-based S6a interface. The MAP protocol has been replaced with an AAA/DIAMETER application (TS 29.272 [8]), which we informally call MAP/IP. Figure 7.15 illustrates the EPS-AV distribution. The requesting MME specifies the type of AV it wants to be returned in the `Network Type` parameter. It is set to `E-UTRAN` for EPS-AKA. The request must also explicitly include the PLMN ID of the requesting MME. The PLMN ID is defined in TS 23.003 [4] and it consists of a Mobile Country Code (MCC) and a Mobile Network Code (MNC). The HSS constructs the AV according to the requested network type, setting the $AUTN.AMF$ separation bit accordingly. For the EPS-AV case, the HSS also uses the PLMN ID as input when deriving the K_{ASME} from the initial (CK, IK) key pair.

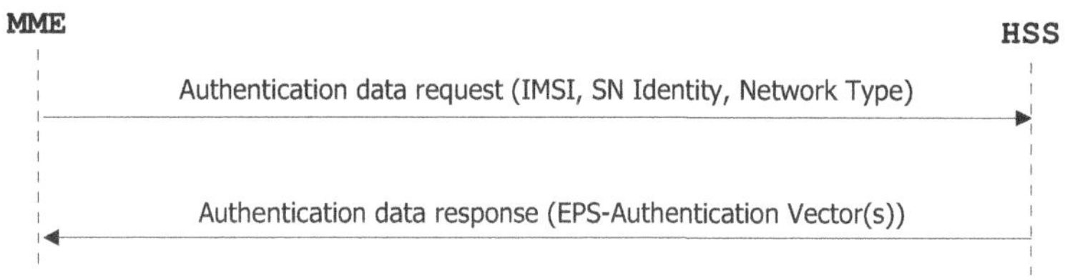

Figure 7.15 Distribution of EPS-AV from HSS to MME

The forwarding messages, the *`Authentication data request`* and the *`Authentication data response`*, must be authenticated (origin- and message authentication) and confidentiality protected. There is no explicit requirement to secure the S6a interface and the MAP/IP protocol, but DIAMETER is designed to use IPsec or TLS. And, clearly, it is possible to use the NDS/IP [13] framework for protecting S6a/DIAMETER. This is strongly recommended.

7.3.6.2 EPS Authentication

The EPS AKA consists of the same challenge as for UMTS AKA, with the exception that one now includes a *key set identifier* (eKSI) in the challenge data. There are two possible eKSI types: The KSI_{ASME} or the KSI_{SGSN}.

The KSI_{ASME} is used to indicate a native EPS security context while the KSI_{SGSN} is used for mapped security contexts. Mapped contexts are used during relocation from UTRAN to E-UTRAN; the MME derives the $K_{ASME'}$ from the CK, IK that it got from the UTRAN (SGSN). The eKSI identifiers are stored with the security context such that one always knows whether the K_{ASME} originated with EPS-AV or is mapped from a UMTS AV. Another difference is that the already mentioned separation bit (bit 0) in AMF, is set to **1** for the EPS-AKA challenge. The EPS authentication sequence, see Figure 7.16, is otherwise identical to its UMTS counterpart. This means that all the UMTS AKA functions $(f1, f2, f3, f4, f5, f1*, f5*)$ are retained.

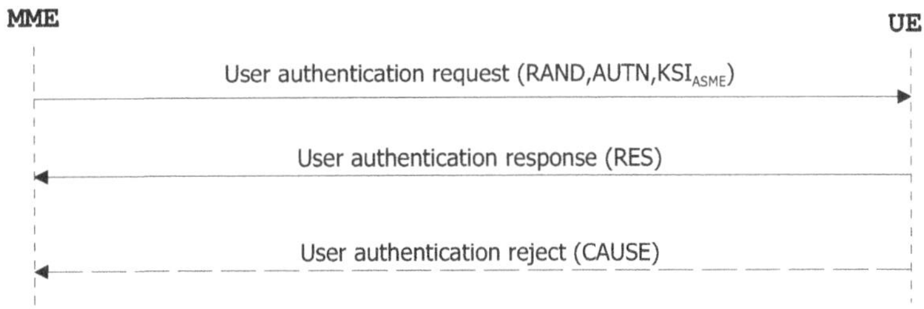

Figure 7.16 EPS Authentication

7.3.6.3 Key Derivation

All EPS keys, including the master key K_{ASME}, are derived by a generic key derivation function (KDF). The KDF is defined in Annex A in TS 33.401 [16] and it is based on HMAC-SHA-256. This KDF is known as "KDF 1" in LTE.

$$derived\_key = HMAC{-}SHA{-}256(Key, S)$$

All the derived keys are initially 256-bit wide, but the session keys are truncated to 128-bit before they are used. For the K_{ASME} derivation the key is $CK||IK$, the string S includes the VPLMN-ID and the anonymized sequence number $(SQN \oplus AK)$.

7.3.6.4 Derived Keys and Key Material

The following keys are defined:

- K_{eNB} – Root key for the `AS Security Context`
  This key is derived by the ME and the eNB when the UE is connected. It is used for deriving the K_{UPenc} and (K_{RRCint}, K_{RRCenc}) keys.

- K_{UPenc} – User Plane
  Used for data confidentiality protection of user plane data over the LTE-Uu interface. There is no corresponding key for data integrity.
- K_{RRCint} and K_{RRCenc} – Keys for protection of the RRC protocol
- K_{NASint} and K_{NASenc} – Keys for protection of the NAS protocol
  These keys are derived by the ME and MME from K_{ASME}.

There is additionally a so-called next hop (NH) parameter/key which is used for chaining of AS contexts during handover.

7.3.7 Weaknesses, Shortcomings and Omissions

When it comes to data privacy, it is fair to say that what LTE does it does well. However, there are some problem areas. The user plane security offered by the `EPS Security Context` only covers the LTE-Uu interface (ME-eNB). The eNBs are widely distributed and one really needs back haul protection. The EPS security architecture strongly recommends protection for the eNB-SGW connection, which would be realized by having NDS/IP [13] protection for the S1 interface. However, S1 protection is not mandatory and many operators do not provide NDS/IP protection (or any other protection for that matter) for this interface.

With respect to identity- and location privacy, the EPS security architecture essentially provides the same (lack) of protection. The TMSI has been replaced by the M-TMSI (in GUTI), but the identity presentation and the use of a temporary identity follows the same pattern and has the same problems as one had for GSM/GPRS and UMTS. Thus, LTE can only offer weak (passive) identity- and location privacy, and obviously credible untraceability is also not possible.

7.4 Brief Summary of Subscriber Privacy in the 3GPP Systems

Data privacy in the mobile networks is based on the data confidentiality protection offered by the access security architecture.

7.4.1 Data Privacy in GSM

Data privacy in GSM is based on the protection provided by the A5 ciphers. The coverage is for the over-the-air interface only (`MS⇆BTS`). This is not sufficient unless the communication to/from the BTS is also protected. There exists no well defined and standardized security for the BTS connections towards the BSC.

The A5/1 algorithm is no longer secure or adequate. Furthermore, the input key, K_C, may be derived from weak algorithms (A3/A8). This problem also affects the otherwise acceptable A5/3 algorithm. The only really acceptable solution is to use the UMTS AKA protocol to derive keys, and then use the A5/4 algorithm for data privacy. However, the A5/4 algorithm is not currently available and even it cannot help with inadequate coverage.

The only conclusion then is that GSM cannot provide credible data privacy. That is, it may be acceptable for some uses, but it will not withstand even quite weak privacy intruders.

7.4.2 Data Privacy in GPRS

Data privacy in GPRS is based on the protection provided by the GEA ciphers. The coverage is for the `MS⇆SGSN` connection, which covers all of the access network and into the core network. While somewhat problematic from a systems point of view, it is actually quite good from a data privacy point of view.

The GEA ciphers, with the exception of GEA4, are all controlled by a 64-bit key. This is not fully adequate anymore. It may not turn out to be a large practical problem just yet, but it certainly isn't future proof. Otherwise the GEA ciphers have stood their ground well this far, but we advocate against GEA (GEA1) since it internally is based on a 54-bit key space.

The GEA4 algorithm is the only one which could be said to be future proof. It requires the UMTS AKA protocol to derive keys, but subscriber should be migrated to UICC/USIM anyway. The biggest problem with the GEA4 algorithm is of course that it currently isn't available.

7.4.3 Data Privacy in UMTS

Data privacy in UMTS is based on the protection provided by the KAUSMI and SNOW-3G ciphers though the $f8$ cipher interface. The coverage is for the `MS⇆RNC` connection. For a conventional RNC this is unproblematic, as it would be a major node located in a secure environment (often co-located with CN servers). Technology progress have made it possible to integrate RNC functionality with the NodeB, and this is a concept which operationally may be seen as useful for home cells (femto-cells). In such a case, connections to the VLR/SGSN must be secured and the RNC itself must be physically protected.

The $f8$-based ciphers are all controlled by a 128-bit key. This is fully adequate. The KASUMI cipher, which internally is a block cipher, has a block length of 64-bit. This means that KASUMI isn't fully future proof. SNOW-3G is a 128-bit design in all respects and so we'd advocate that operators use SNOW-3G as their preferred cipher function.

7.4.4 Data Privacy in LTE

Data privacy in LTE is based on the protection provided by the SNOW-3G and AES ciphers though the EEA cipher interface. For the Chinese market one also has the ZUC cipher suite. The coverage is for the `MS⇆eNB` connection. This is by itself insufficient. It is essential that the connections from the eNB towards the MME (S1-MME interface) and towards the SGSN (S1-U) be protected too. The E-UTRAN network is assumed to be mesh connected and so we also need to ensure that inter-eNB connections are protected. The security architecture standard (TS 33.401 [16]) does indeed strongly advocate protection of these interfaces and the IPsec-based NDS/IP standard [13] is the preferred solution. The NDS/IP protection will set up and keys negotiated on a per node/device basis, and the connections will be independent of any particular subscriber. Normally, one would advocate separating the user plane, the control plane and the O&M plane traffic into different protected connection.

The EEA-based ciphers are all controlled by a 128-bit key and they are all 128-bit designs. This is, of course, fully adequate. The IPsec algorithms allowed in NDS/IP are also quite adequate for the purpose. Configured properly, the LTE network should be able to provide adequately protected connections within the E-UTRAN and towards the LTE core network.

8

Future Cellular Systems and Enhanced Subscriber Privacy

8.1 Background

There are several possible ways to improve subscriber privacy from what is provided in the 3GPP systems. Then there are the regulatory requirements. Privacy would have to be revokable or it must at least be possible to turn off privacy provisioning to fulfill Lawful Interception requirements. Emergency call requirements must also be adhered to. And even large-scale non-user specific "anti-terror" schemes like the UE Data Retention Directive (DRD) must be fit in within the privacy scheme.

Thus, it will be hard to provide bullet-proof privacy, but this shouldn't stop us from trying to have credible privacy schemes that can fend off the most common privacy invasions. To stop law enforcement agencies and the likes to unduly invade our privacy will require political and legislative solutions; One cannot rely on technical solutions to provide both strong personal privacy and at the same time allowing LI/DRD schemes.

The P3AKA protocol presented in this chapter is an attempt at designing a better alternative to the current 3GPP AKA protocol. It is in no way a complete or concrete proposal, but it should be able to demonstrate how both security and privacy could be improved upon relative to the 3GPP protocols. Having said that, it is clear that proposals for structural design change to the identity presentation scheme currently used in the 3GPP architecture are highly unlikely to be accepted. The implementation cost would be huge, the backwards compatibility issues even bigger and the operational costs would be massive. The P3AKA protocol is not a new design per se, but is rather part of a family of protocols stemming back from the PE3WAKA protocol [169] and the PP3WAKA protocol [179], both from 2005.

8.2 Privacy Enhanced Authentication- and Key Agreement

8.2.1 The Principal Entities

Our model has three principal entities:

- **User Entity (UE):** The mobile device, including a security module (typically a smartcard). The UE has a subscription with a Home Entity.
- **Home Entity (HE):** The HE manages global mobility, UE subscription data, UE location data and UE service charging.
- **Serving Network (SN):** The SN operates an access network. A SN will permit a UE to roam onto its network provided the HE and the SN have a roaming agreement. The SN handles local mobility.

Cellular service providers may own both SN and HE networks, but we will assume that HE and SN are managed by separate administrative entities.

8.2.2 Location/Identity Privacy

The subscriber identity- and location privacy protection in the 3GPP systems are inadequate. One cannot avoid situations when the permanent identity (IMSI) is exposed over the radio interface and the TMSI/GUTI-based temporary identity scheme cannot protect against active attacks. Furthermore, the standards fail to recognize the need for privacy protection from the cellular network entities. It is unnecessary and naive to allow all visited networks to know the true identity of a subscriber. One should also ensure that the home network only knows what it has to know. As documented in [121, 168, 221], the UE should not need to expose both its location and identity to the SN and the HE. We have isolated the following requirements:

- **Intruder:** No Intruder should be allowed to learn the identity or the location of a UE. The protection must be effective against both passive and active attacks.
- **HE:** The HE assigns the permanent UE identity. The HE will necessarily know the SN identity and it may need to know the SN server area where the UE is located. However, there should not be any operative reason for the HE to know the precise UE location.
- **SN**: The SN will, due to radio signal measurements, know approximately where the UE is located. Emergency requirements (E112/E911) dictate that the SN must be able to determine the approximate UE position. However, the SN need not actually know the permanent UE identity. The principal SN requirement is that the HE recognizes the

UE, and that it accepts charging for the UE. Consequently, a so-called linkable pseudonymous UE alias identity will suffice.

- **Lawful Interception (LI):** LI capabilities are mandatory requirement for public cellular service operators. Identity- and location privacy requirement will not be above this requirement.

8.2.3 Home Control

The 3GPP AKA protocols are logically off-line with respect to the HE. Indeed, in the 3GPP AKA protocols there is no evidence presented to the HE that the UE is actually presented in the claimed SN network at all. Authentication is delegated to the SN, and this is clearly problematic since HE then has no choice but to accept SN claims regarding the UE service consumption. This requires a lot of HE trust in the SN. A HE typically has a large number of roaming partners (>100) and that level of "blind" trust probably cannot be justified. The HE needs some level of control and enforcement. This makes improved home control a clear requirement.

8.2.4 3-Way Online Authentication

There are three principal parties in our model. The standard off-line 3GPP AKA protocols do not capture this. Neither do they clearly distinguish between the HE and the SN [10, 16, 167], instead referring to both as the "network". To improve home control while retaining SN control, an online 3-way AKA protocol is required. Home control requirements may be at odds with subscriber privacy requirements, but both must be catered for.

8.2.5 Context Binding and Context Expiry

The security context should have limited validity. Protocols like IPsec [163] have exposure restrictions with respect to protection usage (KByte/packets) and lifetime (seconds). For a mobile subscriber, one can additionally extend the exposure control to a geo-spatial dimension. The spatial resolution must be of reasonable granularity. There needs to be a trade-off between signaling performance/workload and privacy aspects. A useful compromise between exposure control and performance may be to assign the spatial binding to the SN server area. This is roughly analogous to binding the context to a SGSN/VLR or MME area. The MME Group area may be too large. Binding to RNC/BSC areas may or may not be too small. The area definitions are somewhat contentious and they may be problematic for femto cells (home basestation), where technological advances already have allowed RNC/BSC functionality to be moved into the basestation equipment.

8.2.6 Principals and Trust

We have the following security trust-relationships:

- **UE–HE:** This relationship is defined by a subscription contract. The HE assigns the permanent UE identity and the long-term security credentials. Security wise, the HE has jurisdiction over the UE.
- **HE–SN:** This relationship is based on legally binding roaming agreements. We assume these relationships to be reasonably long-term and we assume limited mutual trust.
- **SN–UE:** There is no *a priori* SN–UE relationship. We assume transitive trust-relationships and this permits (on-demand) establishment of the SN-UE relationship. Transitive trust is indirect and we assume it to be weaker than direct trust. The SN-UE trust relationship is limited in scope and time to the roaming event.

Transitive trust is not a given. In particular, when the relationship is uneven with regard to needs, detection capabilities and enforcement it is probably not a good idea to assume transitive trust. A subscriber is in a very different position from that of the SN or HE. Asymmetries do not automatically prevent trust, but will impose limits to the trust. Of course, there is also a larger context. The HE/SN are subject to regulatory controls, imposed by the licence they operate under, and all parties are subject to national laws.

Trust with regard to privacy is different from security trust. From the UE perspective the HE is only semi-trusted and the SN even less so. In line with the *Privacy by Design* concept (Sec.1.5), the access security procedures should, whenever possible, aim at concealing UE privacy sensitive data from the HE and from the SN.

8.2.7 The Dolev-Yao Intruder

We assume that the Dolev-Yao (DY) [81] intruder model applies. That is, all that is exposed will and can be used to the full extent by the intruder. The DY intruder cannot physically corrupt principal entities, use social engineering strategies or break cryptographic primitives. However, the DY intruder can (and will) break cryptographic protocols that use cryptographic primitives improperly. Section 3.2.4 also discusses the the DY intruder and the privacy intruder.

8.2.8 The Over-the-Air Intruder

The DY intruder is very powerful and it's not likely that an actual adversary is equally powerful. Proposals for less powerful intruder models exists, and there are concrete suggestions for *Over-the-Air*(OtA) models in which the intruder may be able to read all data, but will generally not be able to prevent genuine messages from reaching the intended recipient. That is, the OtA intruder may be able to corrupt messages, but it cannot guarantee to corrupt all messages. The OtA intruder can inject messages at will and it can be a Man-in-the-Middle agent.

The OtA model can be considered a subset of the DY intruder model, but formal verification research has [234] shown that it is not necessarily easy to demonstrate that a "weaker model" is indeed "weaker". Furthermore, it is often modeling wise much easier to assume the strongest intruder. With this in mind, we assume the full DY intruder to be present in the environment.

8.2.9 Simplified System Model

In a real-world cellular system, there are a relatively large number of system interfaces. In our simplified system model, we only attempt to captures the direct interfaces between the security principals (Fig.8.1). The A-interface is the over-the-air interface between the UE and the SN. It covers common channels and dedicated channels that need to be protected. The common channels are public and unprotected and are generally being used for beacon signaling and during initial access setup. During the setup phase, the A-interface may be severely bandwidth restricted. The B-interface between the SN and the HE is a high capacity interface, which we assume to be permanently available, to have redundant capacity and with fixed propagation delays. We also assume that the HE and SN have pre-established a secure connection over the B-interface. The HE-SN protection is established independent of any subscriber activity and depends only on security credentials decided by the roaming agreement.

8.2.10 Protocol Integration and Round-trip Efficiency

8.2.10.1 Sequential Setup

The 3GPP AKA protocols are all one-pass protocols for the successful case. If we take the case for the relevant protocols (UMTS AKA and EPS-AKA), this is achieved at the cost of having the sequence number replay-protection mechanism [10]. However, the apparent "one-pass" efficiency is an illusion

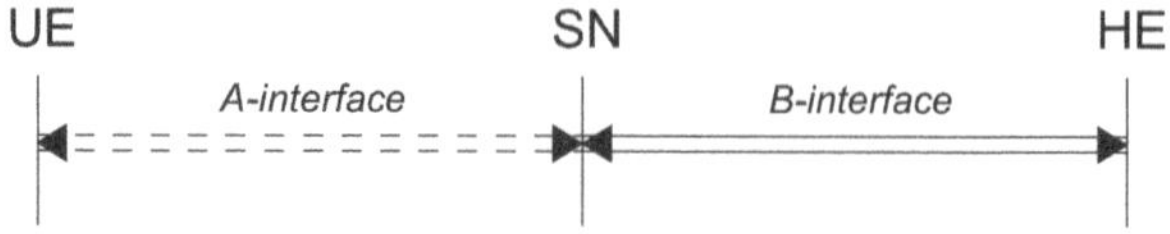

A-Interface: Over-the-air interface, w/limited capasity;
B-Interface: High-capacity authenticated and protected channel.

Figure 8.1 Simplified System Model - Main Interfaces

when the whole set-up signaling is studied [168]. The 3G/4G systems inherited a circuit-switched ISDN-based service model designed for the 2G systems. The 2G control model had to take into account the limitations of the 2G radio-systems and the SS7-based fixed-network signaling. The SS7 protocols suffer under the limitations in the Message Transfer Part (MTP) payload size, which incidentally is also why the short message service (SMS) has the 160 character payload restriction (140 bytes *or* 160 7-bit chars).

Taken together, this has led to a highly sequential set-up. One first completes the Radio-Resource Management (RRM) signaling to establish the physical link. Then, the Mobility Management (MM) procedures are run and finally the Cellular Access Security (CAS) services are run. The number of signaling round-trips to complete a set-up is quite high in the 3GPP systems. This is not satisfactory for a beyond-4G systems with an AIPN service model. To remedy the situation, one must re-design the control signaling. If one combines logically connected sequences one may be able to reduce the total number of round-trips. We do not intend to propose a new control model here, but want to point out that several of the RRM, MM and CAS procedures will/can be triggered by the same physical/external events. Dependencies not withstanding, one can (and should) integrate some of the procedures, and thereby be able to reduce the total number of round-trips. We focus here on the *MM Initial Registration* procedure and the *Authentication and Key Agreement* procedures. In order to get system access the UE must register with the SN (this includes identity presentation) and authenticate itself. The combined procedure must provide identity presentation, registration, and establishment of the security context.

8.2.10.2 Re-keying and Radio Resource Management Events

We note that re-keying events will naturally coincide with RRM events. This allows re-keying to be combined with the RRM event. The LTE/EPS way is to do re-keying for every HO event and this is a sensible approach, and it effectively makes the local/temporary context limited to the cell coverage area. When this happens one may also re-allocate temporary identifiers. Of course, a system synchronized HO in a packet-switched environment is something which in some sense only can occur during packet transmission, but we will assume that it happens for all "active sessions" irrespective of any actual packet transmission at the time. With this in mind, we may advocate that *any* cell change should trigger local security context establishment (with new keys included). One should also, like IPsec does, limit the local scope with respect to usage and passed time. Again, one may follow the LTE/EPS scheme to trigger HO events to re-establish the local/temporary security context.

8.2.11 Proposal for a Security Context Hierarchy

We may classify the security context relations according to its spatial or its temporal coverage. In practice, the two views yield similar structures. One may additionally introduce usage-based context coverage, but in practice only a local session level context will actually contain key material used for communications security. To some extent one may argue that also higher level contexts will suffer from prolonged usage, but it may be problematic to calculate this in our system model. If the keying relationships between the context level have *perfect forward secrecy* properties then the whole argument about higher level context may be slightly moot. We have three basic security context levels in our model (Fig.8.2).

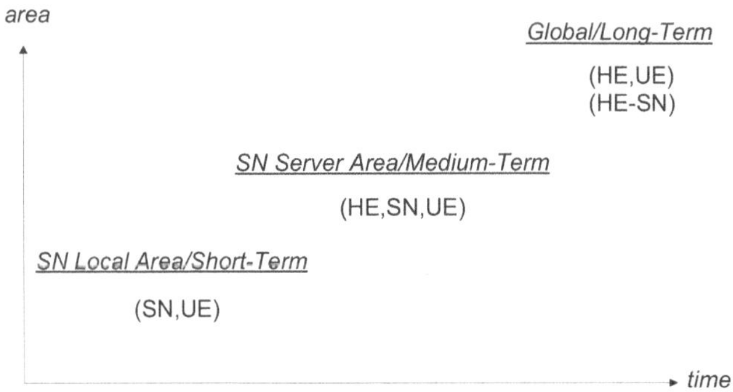

Figure 8.2 Security Context Hierarchy

8.2.11.1 The Long-Term Contexts

In our model, there are two types of long-term contexts.

- **HE–SN:** This context is based on the roaming agreements and will be long-lived. The context will normally apply to the full coverage of the SN network.
- **HE–UE:** This context is defined by the subscription contract. The context contains long-term security credentials. The context is normally valid for all HE provided areas, which in principle is the sum of all HE roaming agreement areas. However, it may also be limited to the home network (pre-paid options etc).

The HE will have a set of HE–SN contexts (typically >100). The SN will likewise have agreements with a large number of HE operators. The HE may serve millions of subscribers (UEs) while the UE will have exactly one HE association.

8.2.11.2 Medium-Term Context

The medium-term context is dynamically established on the basis of long-term contexts. It is categorized by the (UE,SN,HE)-tuple and the context is established in conjunction with the UE registration. The validity of the context is confined by the spatial validity (an Area Code (AC) related to the registration area), a validity period (VP) and possibly a usage count (*K*bytes/packets). The geographical coverage of AC will depend on the network topology. We advocate that the area should be aligned with an area analogous to VLR/SGSN or MME areas in 3G/4G.

The AC may consist of a set of location/paging areas. The validity period, VP, should be long enough to avoid excessive context re-establishments. We advocate that the SN announces the validity period parameter (VP_{SN}). The UE should additionally have a HE-determined maximum (VP_{HE}) value. The context will be canceled when the overall validity period expires ($VP = Min(VP_{HE}, VP_{SN})$). The usage count expiry condition could be negotiated in a similar way.

8.2.11.3 Short-Term Context(s)

The short-term context only applies to the A-interface. The short-term context credentials are derived from the medium-term context. It consists of symmetric-key key material for protection of subscriber related communication. There may be independent keys for uplink/downlink. Likewise, one should have independent keys for control plane and user plane communications. In LTE, we have the case that the NAS security context has different termination points from that of the AS security context. Scenarios like this can be supported by different types of short-term context, and the various types of contexts will likely have different expiry conditions.

8.2.11.4 Context Identifiers

The legal identities, or rather the associated system identifiers, will be part of a long-term context and the scope will be, for our purpose, identical to the spatio-temporal coverage of the long-term context. We propose to define different types of referential identities to the derived context types. That is, we will associate the subscriber with temporary referential identifiers to the medium-term and short-term contexts.

The referential identifiers will be pseudo-random numeric identifiers and they will be cryptographically bounded to the context. The referential identifiers will thereby be linked to the legal identity, but the linkage will be concealed for all but the involved parties on a need-to-know basis.

8.2.12 Identity Presentation and Privacy

Integration of the `Initial Registration` and the `Authentication and Key Agreement` procedures will allow us to save round-trips compared to the standard 3GPP case. This alone is a worthwhile cause. In the 3GPP systems, the `registration` procedure is always initiated by the UE while the AKA procedure is always network initiated. In our combined procedure, we suggest to let it be initiated by the UE. The UE will also be the initiator of our "privacy enhanced 3-way AKA" protocol (P3AKA).

8.2.12.1 The Medium-Term Context Identity

In the 3GPP and 3GPP2 systems, there is a need, occasionally, to present the permanent identity (IMSI) in plaintext over-the-air. In our model, we will completely avoid this.

Instead of having to reveal the permanent subscriber identity, which we denote as $UEID$, we propose to let the UE identify itself with a disposable temporary medium-term context identifier ($MCID$). The $MCID$ shall be a uniformly distributed pseudo-random number decided solely by UE. The $MCID$ value should never repeat for any given UE and it should not be predictable what the next value will be. That is, any particular value shall only be used for exactly one (UE,SN,HE) medium-term context. To let the UE decide the $MCID$ means that the SN may experience $MCID$ collisions. This will occur when two or more UEs choose the same $MCID$ such that their usage overlap in time. A collision event does not have to be a major problem, but the SN will need to reject any context establishment attempt with an existing $MCID$. Needless to say, the $MCID$ collision frequency experienced by an SN server should be very low and the SN needs to monitor the frequency as it may be an incident indication.

To avoid collisions, the $MCID$ value space must be comparatively large. With no bias to the $MCID$ choices, we can use the approximation $p = k/m$, where p is collision probability, k is the maximum number of subscriber within the SN server area and m is the range of the $MCID$ value. To be cautious, let us assume that an SN server can serve one billion simultaneous users ($k = 10^9$). We do not expect a high frequency for establishment of the 3-way medium-term context, but even with the rather unlikely rate of one establishment event every second and a system lifetime of 30 years there will be less than a billion events (10^9) per subscriber. So if we require a collision to occur at most once in the lifetime of a subscriber we have $p = 1/10^9$. To be on the safe side the $MCID$ must then have a range of $m = 10^9 \times 10^9$. To satisfy this requirement, we need the $MCID$ to be stored in a variable with a minimum of 60 bits ($2^{60} > 10^{18}$). This is clearly feasible and in practical terms a 64-bit $MCID$ value will suffice. A cryptographic pseudo-random function will then solve the problem.

$$UE : prf(\cdot) \rightarrow MCID$$

8.2.12.2 Subscriber Privacy and the Medium-Term Context Identity

We ideally want to conceal the permanent UE identity ($UEID$) from both the SN and external intruders. The UE must be able to securely communicate the ($UEID, CID$) association to HE. Data integrity protection is a requirement, and so is data confidentiality in order to avoid exposing the $UEID$. Data confidentiality is also required for the $MCID$ with respect to external parties.

To solve these problems, the UE must privately communicate $UEID$ and $MCID$ to the HE. The SN must know about the $MCID$, and furthermore the HE must corroborate to the SN that $MCID$ is a valid UE identity.

8.2.12.3 Anonymous Tracking and the Need for a Public Identifier

If an identifier is used for a prolonged period an intruder will be able to track the subscriber. The use of pseudo-anonymous identifiers will prevent the intruder from knowing the permanent UE identity, but the intruder could still track the subscriber. The subscriber will be unknown to the intruder, but over time the intruder will learn more and more about the subscriber.

To protect the UE against this type of tracking, we introduce a public identifier associated with the short-term context. We call this the short-term alias identifier ($SAID$), and similarly to the TMSI/GUTI we allow this identifier to be used in plaintext. In our scheme, the SN assigns the $SAID$ in confidentiality protected form. The $SAID$ will be used for paging- and access request purposes. To avoid tracking, the $SAID$ should ideally be assigned for one-time use, but it may be used a limited number of times before being replaced by a new $SAID$. There should be no apparent correlation between the $(UEID, MCID)$ and the $SAID$s, and there should be no apparent correlation between subsequent $SAID$s.

$$SN : prf(\cdot) \rightarrow SAID$$

8.2.13 Use of Identity-Based Encryption

In our P3AKA protocol, we choose to use Identity-Based Encryption (IBE) as our public-key cryptographic algorithm. The interested reader should know that other public-key crypto-systems may also be used for the same over-all purpose. Examples of this may be found in [173].

8.2.13.1 Background

IBE dates back to 1984 when Shamir [227] asked for a public key encryption scheme in which the public key is an arbitrary text string. Shamir proposed a scheme in which the email address could serve as the public key. So when Alice needs to send an encrypted message to Bob she simply uses Bob's email address `bob@ibe.com` as the public key to encrypt the message. There is no need for Alice to obtain Bob's certificate before proceeding and specifically no need for an out-of-band channel to securely obtain the certificate. When Bob receives the encrypted message he contacts the Private Key Generator (PKG) to retrieve the private key. The retrieval process is analogous to private key retrieval from a CA, and it is a prerequisite that Bob and PKG have an *a priori* arrangement with respect to authentication and key protection. Note that the private key can be generated subsequent to the use of the corresponding public key. Since the IBE problem was proposed there have been many attempts to solve it. The first acceptable solution appeared in a paper by Boneh and Franklin [45].

In our scheme, we don't care too much about the cryptography except that we require it to be effective and secure. Thus, we only really care about the functional requirements.

8.2.13.2 The IBE Concept

The Boneh–Franklin IBE scheme is based on three principals and four functions. The principals are Alice, Bob and PKG. The functions are $Setup$, $Extract$, $Encrypt$, and $Decrypt$. The following is a brief description of these functions. Please refer to [44, 45] for a fuller account.

1. $Setup(k) \rightarrow sp, s$
   This function takes a parameter, k, as input. The output is system parameters, sp, and a master-key, s. The system parameters sp does not need to be secret. The master-key s must only be known to the PKG.
2. $Extract(sp, s, ID) \rightarrow d$
   This function take the system parameters, the master-key and an identity string (ID) as input. The output d is the (secret) private key.
3. $Encrypt_{ID,sp}(M) \rightarrow C$
   The function takes as input sp, ID and the message $M \in \mathcal{M}$. The output is the ciphertext C.
4. $Decrypt_{d,sp}(C) \rightarrow M$
   The function takes as input sp, d and the ciphertext $C \in \mathcal{C}$. The output is the plaintext message M.

8.2.13.3 Use of Asymmetric Methods

In the P3AKA protocol the UE is the initiator. The UE must be able to communicate the ($UEID$,$MCID$)-tuple to the HE with full confidentiality protection. To solve this problem, we use an asymmetric confidentiality protection method. We could use any symmetric method, but we prefer to use IBE for this purpose. The main strength of IBE is that it provides us with a way of creating new public keys and starting to use them without a preceding public-key key distribution phase. The construction of the public key allows an elegant spatio-temporal context binding. There is no need for key distribution from the HE to the UE, and keys can be easily updated. The computational capacity of the involved entities is substantial, and we claim that the IBE computational demand is not a compelling argument against use of IBE. By deliberately keeping the HE public keys none-UE specific, the computational burden of generating private keys is contained. Observe that the roles of Bob and PKG will both be played by the HE. This is permissible in our setting since the HE already has security jurisdiction over the UE.

8.2.13.4 Construction of IBE keys

The IBE public key ID will be constructed by the UE in the following way. The public key pair, (ID,d_{ID}), is deliberately specific to the UE.

$$UE : ID := HEID||SNID||AC||VP$$

$$SN : H_1(ID) \rightarrow Q_{ID}$$

$$HE : Extract'(sp, s, Q_{ID}) \rightarrow d_{ID}$$

The ID construction binds the public key to the SN area. The UE knows the HE Identity, while the SN Identity and the area code (AC) are fetched from a broadcast channel. In the Boneh–Franklin IBE scheme [44, 45], the $Extract$ function takes the ID as an input parameter. That is, $Extract$ uses ID to compute $Q_{ID} = H_1(ID) \in \mathbb{G}_1^*$, and it is Q_{ID} that is actually used by the system ($Encrypt$ also uses Q_{ID}). The function H_1 is part of the system parameters, sp and is not assumed to be secret. This permits us to let SN know H_1 and it permits SN to compute Q_{ID} without compromising security. We note that this will not help a rogue SN to obtain information that it would not otherwise see. The purpose of this minor modification to the IBE scheme is to enhance the UE location privacy. With the modification in place, the area code (AC) can be concealed from the HE while achieving the desired local spatial binding.

8.2.14 Key Agreement

8.2.14.1 Prefect Forward Secrecy and the Diffie–Hellman Exchange

We choose to use a Diffie–Hellman (DH) exchange to establish a shared secret for the medium-term context. The DH shared secret (dhs) will be used as the basis for deriving the session key material that is needed in the short-time context(s). In the simplest case, this will amount to one data confidentiality key and one data integrity key. We do not require perfect-forward secrecy for the short-term contexts derived from the medium-term context. We therefore allow re-use of the dhs for generation of short-term contexts between the UE and the SN during the lifetime of the medium-term context. To permit this, we require that the dhs have at least 256 significant bits. This will amount to exchange of DH public keys in the order of 15K bit in size.

8.2.14.2 Unconventional use of the DH Exchange

The channels available during the early setup phase are bandwidth restricted. We cannot expect to be able to carry out the DH exchange with 15K bit public keys over the A-interface. It is possible to save bandwidth by using ECC-DH (which requires only approximately 600 bits pr key), but even this may be too much. Since the HE has security jurisdiction over the UE, we may allow the HE to carry out the DH exchange on behalf of the UE. The HE must then somehow be able to communicate the dhs to the UE after the DH exchange has taken place. Needless to say, the transfer must be fully authenticated and data confidentiality protected. We denote the public-key HE parameter DH_{HE} and the SN parameter DH_{SN}. Details concerning the DH group agreement etc could be part of the HE–SN long-term security context.

8.2.14.3 Symmetric Key Derivation

The symmetric key agreement scheme will use the DH secret (dhs), an identity and an area code as the basis. For the medium-term context, the area code is the AC. For the short-term context, a local area code (LAC) is used. This area code may be the *location/paging area* code. For the sake of simplicity, we denote the symmetric keys simply as mtk and stk according to context; medium-term and short-term respectively. In a real-world system, there may be many different keys like we saw for the LTE case.

$$UE, SN : KDF_{dhs}(MCID, AC) \rightarrow mtk$$

$$UE, SN : KDF_{dhs}(SAID, LAC) \rightarrow stk$$

The key derivation function $KDF(\cdot)$ may be constructed along the lines of the LTE KDF. The mtk (medium-term key) is only used between the UE and SN during execution of P3AKA protocol. The stk denotes the short-term key set. The stk and the associated $SAID$ are valid for the duration of the short-term context.

8.2.15 Entity Authentication

8.2.15.1 HE–UE Authentication

To avoid excessive use of IBE or other asymmetric methods, we also let the UE and HE share an authentication key KA to be used as the basis for the two-way checksum-based challenge-response mechanism. Another HE–UE key, KIC, is used for confidentiality and integrity protection between HE and UE during P3AKA execution. This key is derived from the $MCID$ and $UEID$ and with KA as the basis.

$$UE, HE : KDF_{KA}(UEID, MCID) \rightarrow KIC$$

The pseudo-random challenge data (CH_{UE}, CH_{HE}):

$$UE : prf(\cdot) \rightarrow CH_{UE}$$
$$HE : prf(\cdot) \rightarrow CH_{HE}$$

Response data (RES_{UE}, RES_{HE}):

$$UE, HE : R_{KA}(CH_{UE}, MCID) \rightarrow RES_{UE}$$
$$HE, UE : R_{KA}(CH_{HE}, MCID) \rightarrow RES_{HE}$$

To properly bind the challenge-response to the medium-term context, the $MCID$ is included as an input parameter. The response function $R(\cdot)$ is a keyed one-way function (MAC). The key KA is included in the long-term security context between the UE and HE.

8.2.15.2 HE–SN Authentication

We assume that secure communication over the B-interface has already been establish and that the SN and HE are mutually authenticated. The secure communications over the B-channel is generic to the HE–SN relation and it is not related to any particular subscriber.

8.2.15.3 SN–UE Authentication

To protect its interests, the SN must insist that the UE be authenticated. We shall allow this to mean that SN has assurance that the HE accepts the $MCID$ identifier as a valid UE identity. The SN already knows that HE has security jurisdiction over the UE, and if the HE can assure the SN that it has authenticated the subscriber then SN shall accept $MCID$ as a valid UE identity. This information is conveyed to the SN over the B-interface. The SN and HE have mutually created the DH secret dhs online, and they are both assured that dhs is a fresh shared-secret for the $MCID$ medium-term context. If the UE, which claims the $MCID$ identity, can show proof of possession of dhs, then the SN shall be compelled to believe that the UE is the principal that HE has assigned the $MCID$ identity to. Likewise, the UE constructed the $MCID$ and knows that it is fresh. Given that the UE has received the dhs from the HE in an authenticated and encrypted form, the UE has assurance that the HE intended the dhs to be the context secret for the $MCID$ context. Given the trust relationships and the HE jurisdiction over the UE, the UE will accept the SN ($SNID$) as an authenticated entity provided it can show proof of possession of the dhs. Proof of possession of dhs will be demonstrated by the use of the symmetric keys (mtk), which is derived from dhs.

8.3 The P3AKA protocol

8.3.1 Main Objectives

The main objectives of the P3AKA protocol is as follows:

- **Authentication**
  The protocol must provide mutual entity authentication between (UE–HE) and (UE–SN).
- **Local Re-keying**
  The protocol must allow local re-keying over the A-interface.
- **External Subscriber Privacy**
  The protocol must be able to prevent an external intruder from learning the permanent UE identity and/or location and from tracking an anonymous UE.
- **Internal Subscriber Privacy (SN)**
  The protocol must be able to prevent the SN from learning the UE identity. SN will, for radio transmission related reasons, invariably know the approximate UE position.
- **Internal Subscriber Privacy (HE)**
  The protocol must not allow the HE to learn the precise UE position. The HE assigned the permanent UE identity.

A premise of the P3AKA protocol is that the HE and the SN do not collaborate to deceive the UE. This is also the premise for all of the 3GPP-based AKA protocols. Indeed, given that the HE has security jurisdiction over the HE, we must require the HE to act honestly towards the UE.

8.3.2 Lawful Interception Requirement

We assume that future public cellular systems must comply with Lawful Interception (LI) requirements. It must be possible to correlate the HE and SN information to permit a Law Enforcing Agency (LEA) to determine both the identity and position for any given UE. The P3AKA protocol permits the data to be revealed given HE and SN cooperation.

8.3.3 Outline of the P3AKA Protocol

We now present and examine the P3AKA protocol in more detail. An outline is depicted in Fig.8.3. The keys used for B-channel protection are not part of the P3AKA protocol.

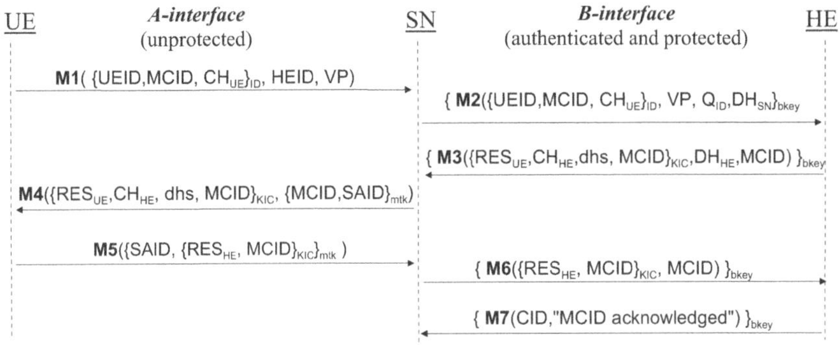

Figure 8.3 The P3AKA protocol

8.3.4 The P3AKA Protocol in Extended Alice–Bob Notation

1. Message M1
   UE constructs the validity period (VP), the HE public key (ID) and the medium-term context identity ($MCID$). The UE then generates the challenge-response data (CH_{UE}, RES_{UE}) and the KIC key.
   $A := \{UEID, CID, CH_{UE}\}_{ID}$

   M1: $UE \rightarrow SN : A, HEID, VP$

2. Message M2
   The HE address is derived from $HEID$. SN constructs DH key (DH_{SN}). The area code (AC) is extracted from the radio system.
   $H_1(HEID||SNID||AC||VP) \rightarrow Q_{ID}$

   M2: $SN \rightarrow HE : \{A, VP, Q_{ID}, DH_{SN}\}_{bkey}$

3. Message M3
   HE generates d_{ID} (if needed) and decrypts A. HE now *sees* claimed UE identity ($UEID$) and the context identifier ($MCID$). HE computes the response (RES_{UE}) and the challenge/response ($CH_{HE} - RES_{HE}$). The HE also derives the KIC key. HE generates DH key (DH_{HE}) and computes the shared secret (dhs). $B := \{RES_{UE}, CH_{HE}, dhs, MCID)\}_{KIC}$

   M3: $HE \rightarrow SN : \{B, DH_{HE}, MCID\}_{bkey}$

4. Message M4
   SN decrypts M3 to get B, DH_{HE} and $MCID$. SN computes (dhs) and generates the short-term alias identifier $(SAID)$. SN derives medium-term key mtk and short-term key stk.

   M4: $SN \rightarrow UE : B, \{MCID, SAID\}_{mtk}$

5. Message M5
   UE decrypts B and verifies that HE has accepted $MCID$. UE verifies the response (RES_{UE}) and computes the response (RES_{HE}) to the challenge (CH_{HE}). Based on dhs, $MCID$ and AC the UE generates mtk, which it uses to decrypt the reminder of message M4. The UE sees the $SAID$ and can verify that it is bounded to $MCID$. With dhs, $SAID$ and LAC, the UE generates session keys (stk).

   M5: $UE \rightarrow SN : \{SAID, \{RES_{HE}, MCID\}_{KIC}\}_{mtk}$

6. Message M6
   SN decrypts M5 and verifies the $SAID$ assignment. By now the UE–SN short-term context is available. The SN still needs the final confirmation from HE that $MCID$ is authenticated, but the SN essentially has this confirmation given that UE has demonstrated possession of dhs (which the SN knows to be fresh). The short-term context can be used from now on.

   M6: $SN \rightarrow HE : \{\{RES_{HE}, MCID\}_{KIC}, MCID\}_{bkey}$

7. Message M7
   HE decrypts M6 and decrypts the UE response. HE then verifies the response RES_{HE} and confirms this to the SN.

   M7: $HE \rightarrow SN : \{MCID, "acknowledge"\}_{bkey}$

8. **P3AKA termination**
   SN verifies the HE acknowledge information.

8.4 Analysis of P3AKA protocol

8.4.1 Authentication and Identity Handling

8.4.1.1 Relative Efficiency

The 3GPP AKA protocols are so-called single round-trip protocols. This apparent round-trip efficiency is possible since identity presentation and acknowledge messages are not directly part of the protocol. The 3GPP AKA protocols rely on a sequence number scheme for replay protection. The use of a sequence number scheme is not unproblematic and unless properly configured it may either cause frequent re-synchronization events or it may cause weakened replay protection. Thus, it adds complexity and provides weak replay protection [187]. In the P3AKA protocol we decided to avoid the use of sequence numbers and related schemes.

8.4.1.2 Double Challenge-Response

We also note that the off-line nature of the 3GPP AKA protocols has the consequence that the protocols cannot really provide "mutual" entity authentication. In the P3AKA protocol we provide true mutual entity authentication between the UE and the HE based on a standard double challenge-response scheme. The scheme used is straightforward and we expect the method to be sound. The challenge-response method is based on keyed checksums, where the (UE–HE) shares an authentication key (KA). The inclusion of $MCID$ provides context binding and freshness assurance. The HE and UE will, therefore, be compelled to believe in the authentication and to accept $MCID$ as a valid medium-term context identifier.

8.4.1.3 Trust Relationships and the Diffie–Hellman Secret

The authentication between the UE and SN is somewhat more complex. The identity corroboration is indirect and relies on the transitivity of the SN–HE trust relationship and the fact that HE has security jurisdiction over the UE. The SN participates in generation of the DH shared secret (dhs), which is directly associated with the $MCID$ context identifier. The P3AKA protocol is an online protocol and the B-interface is pre-authenticated and protected. Thus, the SN can safely believe that the dhs is a fresh shared secret. Based on the roaming agreement contract, the SN can justify a belief that HE has jurisdiction over the UE. This includes a belief that HE can communicate securely and privately with the UE. When the SN receives message M5, it has proof that (the presence of alias identity $SAID$) UE possess the shared secret dhs. The SN therefore in a position to accept that $MCID$ is an authenticated identity. It furthermore can support a belief that the HE will accept liability for the entity presenting itself with $MCID$. There will obviously be conditions and limits to the amount of liability, framed in the subscription contract and the roaming agreement, but we shall ignore those aspects here.

8.4.1.4 Honest Principals

One often relies on the assumption that the principals will be honest and act according to "intentions". However, this isn't always so and it is prudent to ask what would happen if the principals are dishonest.

Could the UE and HE carry out masquerade to trick the SN into being confused about the $MCID$ identity? The answer to this question is that the SN should not care. The SN only needs assurance that the HE will accept charging liability for the $MCID$. Thus, there is no point for the HE to "falsely" accept a $MCID$ and thereby permit masquerade. So, if indeed the HE has accepted $MCID$ as identifying one of its subscribers, then the SN shall accept the $MCID$.

Could the UE and SN try to trick the HE? Well, the subscriber certainly can try to do this and it may very well install malware on the device etc. to aid in a con trick. The SN may cooperate (or it may be a "false" SN), but what could be gained from all of this? A "false" SN should never be allowed to set up a secured connection, so we must assume that it is a valid SN. The SN may be corrupt in some way and this may apply to the subscriber too. With respect to the HE there would basically be a couple of "tricks" to be played: Charge too much or charge too little, charge the wrong subscriber. We note that the SN can charge the HE falsely, and that this ability isn't really influenced by the AKA protocol very much. So it would seem that the SN cannot gain much from this anyway. The subscriber would stand to benefit if it can trick the SN/HE into not charging it. The subscriber may be able to corrupt the user identity module, but it would stand to gain very much. That is, unless the subscriber is able to steal someone else's security credentials. In that case, fraud is possible. This case does not actually involve a dishonest principal per se, but rather an intruder with the ability to physically attack and extract the security credentials.

What about the HE trying to trick the UE? The question is moot in the sense that the HE has jurisdiction over the UE. The HE can cheat on the UE at will, but this capability is independent of the AKA protocol.

8.4.2 Key Exchange

Tripartite Diffie–Hellman methods have been devised, like the one developed by Joux [158]. A tripartite DH scheme could have been useful as the basis for the P3AKA protocol, but restrictions on the A-interface makes it hard to justify this. Furthermore, the physical communications topology is such that all HE–UE communications would necessarily be over links controlled by the SN. This is undesirable and it may even be problematic. The key derivation method as presented here is not new or novel. We are confident that the method used for production of session key material is sound. Subsequent local key exchanges would use the same mechanism as shown in the P3AKA protocol (mtk and dhs are available for this purpose).

8.4.3 Spatio-Temporal Context Binding

The medium-term security context is tied to the context identity ($MCID$) and the ID key. The ID key binds the context to the specific HE and SN. The context is confined spatially by the area code (AC) and temporally by the validity period (VP). Both the SN and UE must verify that the context is valid before using it. The short-term context is tied to the alias identity ($SAID$), which is associated with the ($MCID$) and bounded to the local area code (LAC). A session may exceed the validity (spatially and/or temporally) of a short-term context. Re-keying is then necessary, and in our scheme this is connected to assignment of the alias identity ($SAID$). Local re-keying is then triggered by $SAID$ assignment. A session may extend beyond the validity of a medium-term context. The P3AKA protocol must then be re-run to re-establish the context. The medium-term temporal context expiry is also observed by the HE. Subsequent to a VP expiry, the HE will no longer route data towards the UE unless it re-registers.

8.4.4 Location- and Identity Privacy

The stated goal was to prevent an outsider (intruder) from learning the $UEID$ trough eavesdropping or manipulation of over-the-air data. The $UEID$ is never transmitted in clear. Provided that the crypto-primitives are safe and secure, and that the intruder cannot gain access to the private key (d_{ID}), we shall consider this requirement to be fulfilled. We also required that the SN never learn the permanent UE identity. We observe that the $UEID$ is confidentiality protected and that the private key is not available to the SN. This ensures that the SN will never learn the $UEID$. Furthermore, the HE should not learn the UE location. The HE will necessarily know in which roaming network the UE is located. The spatial binding of the area code (AC) to the medium-term context could have been problematic, but the use of the IBE hash H_1 means that the HE will not actually see the AC. The HE will, nevertheless, have assurance that the SN claimed area code is verified by the UE. The HE will therefore have a weak measure of spatial control. There are ways to extend the spatial control if HE home control policies dictates it [177, 178]. A strong measure of spatial home control while still observing UE location privacy is also possible by using Secure Multi-party

Computation methods [179], but this scheme may be too costly to operate to justify the enhanced home control. Prevention of subscriber tracking was also a privacy goal. We note that from an information theoretic perspective one must prove the absence of any kind of correlation between each access (including usage patterns etc.) to prevent tracking. This is outside the scope of the P3AKA protocol. Instead, we consider this goal to be satisfied provided that the short-term sessions are sufficiently short, i.e. that new alias identities ($SAID$) are used for separate access requests. The lifetime of active sessions must also be contained to fulfill this goal.

8.4.5 Communication Aspects

8.4.5.1 A-interface

A concern for the P3AKA protocol has been the capacity of the (radio) common channels of the A-interface. The element bit-sizes given below is only an estimate:

- **M1:** The M1 message contains one IBE (ECC) encrypted block and the $HEID$ and VP strings. The identity $HEID$ is expected to be 128-bit wide, and the validity period VP may also consume 128-bit.
- **M4:** The M4 message consists of return data from HE (B), encrypted with symmetric methods (we assume 128-bit blocksize, and an added cryptographic checksum (64-128 bit)). If we assume that RES_{HE} and $MCID$ requires 64-bit storage, we have that we need 640 bit for block B. The remaining data from SN should fit in 256 bits of storage.
- **M5:** M5 contains response data to the HE and the $SAID$. It should require no more than 512 bits of storage.

As indicated above, the message all require less than 1 kbit. Even systems with severe bandwidth restrictions during the set-up phase should be able to accommodate the modest requirements of the P3AKA protocol.

8.4.5.2 B-interface

The B-interface is a fixed line interface. We do not foresee any capacity problems here.

8.4.6 Computational Aspects

The total computational demand of the P3AKA protocol is significantly higher than for the 3GPP symmetric-only AKA protocols. However, modern smart-cards are powerful devices and have dedicated crypto hardware. The core network nodes are powerful servers and we cannot see that being a problem either. All in all, there is no compelling reason why the P3AKA protocol cannot be supported by a next generation wireless/cellular system.

8.4.7 Denial-of-Service (DoS)

The P3AKA protocol does not provide explicit DoS protection. The P3AKA protocol operates in an environment where it is very easy for an adversary to carry out access-denial DoS attacks simply by disrupting the radio transmission. Access-denial attacks are local in nature and since they do not scale we have deliberately not tried to avert this type of attack in the P3AKA protocol. To limit computational DoS attacks, we suggest that the SN restricts the arrival rate of P3AKA invocation per access point. The HE, likewise, may limit the number of simultaneous P3AKA sessions from any given SN. Together, this will prevent a computational DoS attack from scaling.

8.4.8 Round-trips

Our claim was that the P3AKA protocol be at least as efficient as the 3GPP schemes. This amounts to assessing the cost of a 3GPP location updating sequence including the 3GPP AKA protocol and comparing it to the P3AKA round-trip cost (ref.: TS 23.108 (ch.7.3.1) [5]):

- UE→SN: *Location Updating Request*
- SN→UE: *Authentication Request*
- UE→SN: *Authentication Response*
- SN→UE: *Location Updating Accept*

The TS 23.108 specification only covers the UE–SN communication. The SN-HE part can be found in TS 29.002 [7], and it constitutes two separate request-reply sequences where the SN first fetches the subscriber information and then the security credentials. In P3AKA, the subscriber information will be forwarded in parallel with message M3. The P3AKA protocol performs slightly better than the 3GPP scheme on the UE–SN interface. Note that message M4 serves as location registration confirmation. On the SN–HE interface, the P3AKA protocol requires two passes. The 3GPP scheme also requires two passes, but they may be executed in parallel. We note that the UE–SN context is potentially operative after SN reception of message M5. The SN still wants HE confirmation, but it now has indirect $MCID$ confirmation. It is, therefore, safe for the SN to activate the short-term context.

With this in mind, we can defend the claim that P3AKA is at least as efficient as the comparable 3GPP sequences.

8.5 Alternative Approaches and Related Research

An alternative scheme using group pseudonyms is investigated in [164]. This scheme is GSM specific and we do not consider it to be directly relevant for a beyond-3G protocol.

In [107], the authors investigate the use of MIXes. However, MIXes typically have indeterministic delays and one would need a set of MIXes to achieve sufficient privacy. MIXes can certainly be designed to work in a real-time setting, but then traffic analysis using timing characteristics would likely be very effective.

Most of our cellular/wireless privacy requirements have been identified previously, and Asokan's work from 1994 are still valid in this respect [32]. Another old, yet valid work, in this area is found in [221]. However, the solutions presented in those papers do not offer sufficient home control. In fact, it was a design goal to avoid the global round-trip back to the HE. When we compare them with our scheme, we find that our security context hierarchy is more flexible and will afford better home control. Our scheme is also very different from [221] when it comes to identity management. Another paper that deals with user privacy and wireless authentication is aptly titled paper "Untraceable Mobility or How to Travel Incognito" [34]. Again, we find that many of the requirements are similar to our requirements. The solutions proposed are quite different from our approach. A direct comparison between our P3AKA and the suggested protocols in [34] would be unfair since they deliberately considered low-cost solutions while we aim at a complete solution with a redesigned identity scheme etc. Finally, many of the requirements we have arrived at are also found in [121]. The solution suggested in [121] does not take into account the possibility of integrating security setup with MM procedures, and it has a very different approach to identity management.

8.6 Summary

We have presented and analyzed the requirements for a privacy enhanced authentication and key agreement protocol for use in beyond-3G public cellular systems. The proposed P3AKA protocol is capable of substantially improved user location/identity privacy compared to the 3GPP AKA protocols. The

P3AKA protocol provides enhanced privacy from eavesdropping and manipulation by external parties and it provides a measure of user location/identity privacy with respect to the SN/VPLMN and the HE/HPLMN. To achieve this, the subscriber identity presentation and the initial registration procedures had to be modified from the procedures used in 3GPP systems. However, this change also allowed performance improvements by integration of signaling procedures that is anyway triggered by the same physical events. The P3AKA protocol also recognizes that all three principals should participate in the security context establishment and that all parties should be online during the establishment procedure.

Finally, the P3AKA protocol provides enhanced key distribution relative to the 2G/3G AKA schemes. In the P3AKA protocol, the SN/HPLMN participates actively in creating the security credentials for session key derivation. Contrast this with 2G/3G scheme in which the HE/HPLMN unilaterally decides the session keys. The difference is less distinct when one compares with LTE/LTE-Advanced, and the differences between P3AKA and EPS–AKA in this respect are mostly due the fact that P3AKA is an example protocol while EPS-AKA is a deployed protocol. The additional complexity found in the EPS key hierarchy is more a symptom of complexities in the EPS architecture than a symptom of bad design in EPS–AKA as such.

We have not provided formal proofs of the security properties of the P3AKA protocol and until the protocol has been subjected to formal analysis and verification one must remain skeptical to the claims made for the protocol. Formal proof isn't, however, a silver bullet and one should be vary about claims about correctness when it comes to formal proofs. By their very nature that can only capture properties about the model within the specific formalism. Formal proof tools are very useful, but primarily as a design tool and to avoid embarrassing errors [124]. This notwithstanding, the generic approach behind the P3AKA protocol seems to be a promising approach for improved privacy in authentication protocols.

Part III

Ubiquitous Devices and Services

9

Privacy and Sensor Networks

In this chapter, we discuss why and how deployment of wireless sensor networks creates privacy concerns and consider some privacy-preserving solutions proposed recently.

9.1 Wireless Sensor Networks

Wireless sensor networks (WSNs) are usually deployed to support distributed interaction with the physical environment through measuring and aggregating data. Such networks naturally are a part of a broad range of applications that involve system monitoring and information tracking for managing critical assets. As an example, consider the task of Structural Health Monitoring (SHM) concerned with monitoring the integrity of civil and military structures. Another example is a WSN for monitoring of the vital signs parameters of patients in a metropolitan environment, where situations of data loss, corrupted or delayed data can have implications of life and death. The last example illustrates a case when privacy protection measures should be implemented. Managing critical assets requires high trustworthiness guarantees, which may be difficult to achieve in WSNs since, most security protocols and architectures have been primarily designed for wired networks and in comparison only a little work is available on addressing security efficiently in WSNs. One way to increase trustworthiness of monitoring data is to use cryptographic protection to avoid open transmission of critical information. However, sensors are small devices with very limited computational and energy resources, while most data protection protocols and architectures have been primarily designed for wired networks, which have more substantial computational resources.

Typically, a WSN consists of a large number of different types of sensors that are able to monitor a wide variety of ambient conditions such as temperature, humidity, vehicular movement, pressure, noise levels, etc. The low cost of sensors makes it possible to deploy a large number of them to perform reliable monitoring and data aggregation. However, the low cost of sensors also leads to severe resource constraints such as limited battery power, memory and low computation capability, and these constraints in turn introduce major obstacles in the implementation of traditional computer security and privacy protection approaches (such as based on computation-demanding cryptography or communication-demanding protocols) in a WSN. Moreover, the open nature and unattended operation of WSNs make the security defenses even more difficult. Another feature of WSNs is that sensor nodes can be either inaccessible, or easily accessible and unprotected. In the first case, ongoing sensor maintenance may be problematic. In the second case, the data received or routed though such sensor may be tampered and therefore not truthful. Privacy in WSNs encompasses several different aspects as outlined below.

- Data privacy: Only authorized parties should be able to recover and understand data delivered by sensors. It can be partly achieved by security techniques.
- Access privacy: preventing WSN from learning what data the users actually have accessed or learning their access patterns.
- Location privacy of sensors: Only authorized parties should be able to discover the location and activity of sensors.
- Privacy-preserving data monitoring: sensors will not need to disclose the measurement values to the monitor be able to discover event of interest in the stream.

9.2 Motivating example: Wireless Health Monitoring

As a motivating example, we consider a WSN for monitoring vital signs parameters from patients in a metropolitan area [254]. Such network includes body sensors communicating with a receiver unit, e.g., a Hand Held Device (HHD) carried by a patient, which in turn can use another wireless hoop (for example, 3G telecommunication solution) in order to transfer data to a central base station. The HHDs can be bypassed when a patient is at home and the body sensors can transmit directly to a home-installed stationary wireless sensor HUB instead of using a wearable HHD. Depending on the actual situation the sensors should be provided with extra bandwidth. For example, when patient condition is normal, body sensors can report their parameters infrequently. In case of an abnormal situation, the sensors should automatically increase the data transmission rate and report health parameters more frequently. Under critical health conditions, the sensor will need maximum bandwidth for near-to-continuous transmission of monitored parameters.

Privacy protection is critical in a wearable sensor network. Security precautions must be taken to ensure correct operation. For example, a body sensor will have to transmit individual unique number (SensorID) which has to be associated with the correct patient (PatientID). Both PatientID and SensorID are required to asses patient condition with respect to the measured vital signs. Sensor readings associated with HHD or HUB can also reveal the geographical location of the patient. Meanwhile, patient location should not be exposed under normal circumstances, while in critical situations both sensor location and the accurate values of the monitored parameters must be provided. For example, consider a patient monitored for his heart condition by a wireless ECG (Electro cardiography) sensor detecting abnormal heart beats and life threatening cardiac activities. If the sensor detects a sudden heart attack (ventricular fibrillation), it is time-critical to start a rescuing procedure. Thus, the sensor should be able to trigger an automatic escalation of the WSN, where this sensor is given a higher priority to ensure reliable data transmission. At the same time, this situation should open protected data to give the rescuing personnel more information about time, place and other relevant data transmitted from the area of accident.

Thus, sensor networks transmit monitoring data via wireless medium and thus are vulnerable with respect to privacy and security. Sensor measurements represent private information about monitored objects, which implies that data transmissions and data flow within and out of sensor network should be protected. At the same time, WSNs should provide efficient authorized access to measured data in order to perform monitoring and event detection. Since sensors have limited battery life-time, low data transmission rates and computational power, traditional privacy protection approaches based on strong encryption cannot be applied directly. From other side, in many cases application of cryptography is not necessary to archive monitoring goals. For example, the different degree of information leakage can be accepted as a result of trade-off between requirements to privacy and performance. In order to reason about such trade-offs, the WSN should utilize application semantics to decide what parameters should be transmitted, what levels of refinement in parameter values (e.g., actual data readings, or an aggregated estimate) should be provided. Automatic escalation algorithms should take this application semantics to choose an appropriate transmission mode, which can be implemented by either transmitting more parameters, or transmitting more refined values of some parameters.

9.3 Access privacy

The traditional access model of WSNs assumes the owner and user of the WSN to be the collector and consumer of sensor readings and trusting each other. However, in reality, owners and users of the sensor network are different. The owners comprise multiple organizations that do not necessarily fully trust each other but need to collaborate for administrative purposes. Because of legal requirements, security and privacy requirements or economical reasons, owner of the network may require that only authorized users may access the sensed data. From other side, users may not be willing to disclose their interests and want to preserve their privacy while accessing the sensed data. The users want to prevent the network owners from learning what data the users actually have accessed or learning access patterns. For instance, an oil company interested in the data of an ocean sensor network may want to hide its network regions of interest from both the owner and other users of the network (who might be potential business competitors) [65]).

In [65], authors consider how to preserve privacy of clients querying a WSN owned by *untrusted* organizations. The authors consider a system where a sensor network is operated by (at least) two servers, and shared among multiple users. Users access the network by sending queries into the network through the server(s). The system is administered by mutually distrusting organizations. The authors propose efficient mechanisms for privately querying sensor networks, that is, enabling users with private access a large-scale wireless sensor network, when access to sensor readings is provided by servers governed by these organizations. They consider two cases: (1) the servers are honest-but-curious, and (2) servers exhibit malicious behavior, for example, as a result of an external attack that gave access to sensor network traffic information.

In [255], authors propose a Distributed Privacy-Preserving Access Control scheme for sensor networks. It is assumed that the network has a single owner but many users want to access data produced by the network. In the proposed scheme, sensor nodes reply only on query from users who have valid tokens which they may purchase from the network owner to achieve privacy preserving user identities and corresponding tokens be unlinkable. It is achieved by use of blind signatures in token generation. To prevent malicious users from reusing tokens, the authors propose a distributed token reuse detection schemes that provide detection without involvement of the base station.

9.4 Location Privacy

The open nature of a WSN makes it relatively easy for an adversary to physically localize both data source (sender) and sink (receiver) by eavesdropping and tracing packet movements. Location privacy in WSNs refers to the ability to protect against such tracing and may be classified into two categories: location privacy of sensor nodes (e.g., data source) and location privacy of sinks (receivers, base stations, etc.). Approaches proposed to protect location privacy in WSNs are designed to protect either against *traffic analysis attacks* (as in [75]) or against *packet-tracing attacks* (as in [151]). Defending against such attacks usually can only partly be cryptography based since, even the adversary cannot analyze the content of packets, she can gathers information by overhearing and following radio communication, following encrypted packets, analyzing traffic volume or time correlations.

9.4.1 Source Location Privacy

Source location privacy means protecting the location of a node reporting the data reading (source) from attackers who are capable performing traffic analysis. Since a source node reports events that in the close proximity when the attacker locating the sending node will help the attacker to locate the reported event. The problem was first illustrated as the Panda-Hunter Game in [160]. In the Panda-Hunter Game, we assume that the large sensor network is deployed to monitor movements of pandas. Each time a panda is sensed, the corresponding sensor node will report this event by sending messages to the sink (the base station). An attacker (the panda hunter) might try to intercept transmitted messages that report a panda observation event and try to localize the panda via localizing the source node. In this case, the hunter will use existing infrastructure to find pandas. (Note that even if the packet content is encrypted, the hunter may still retrieve location-sensitive information from their headers.)

Since the most of existing protocols cannot provide source location privacy without a significant increase in resource consumption, a privacy enhancing technique called *phantom routing* was developed [160] to protect source privacy. The phantom routing delivers each message in two steps: (1) direct the message to a phantom source by means of random or direct walk), and (2) direct the message to the sink by means of either flooding or single path routing (depending on underlying routing protocol). This approach provides protection of the source's location without significant increase in energy overhead.

9.4.2 Receiver Location Privacy

Receiver (base station) is a central player in data collection and a single point of failure. By distorting or destroying receiver, the adversary will make (at least a part of) WSN inoperative and prevent data-gathering.

To prevent the adversary from localizing and physically capturing the receiver by using packet-tracing, a location-privacy routing protocol is proposed in [151]. The idea is to make directions of both incoming and outgoing traffic at a sensor node uniformly distributed by providing path diversity and injecting fake packets. However, the use of path diversity and fake packet injection leads to higher overhead because of longer routing paths and more packet transmissions. Stronger privacy protection requires higher overhead. The proposed approach is designed to protect against packet-tracing attacks.

Since the nodes near the base station usually forward a greater number of packets than remote nodes, an adversary may analyze the traffic patterns (known as traffic analyzes attack) to deduce the location of the base station within the WSN. Traffic analysis attacks in WSNs can be divided into two classes: *rate monitoring attacks* when an adversary moves closer to the nodes having a higher packet sending rate; and *time correlation attacks*, when an adversary observes the correlation in sending time between a node and its neighboring node. To prevent such traffic analysis attacks, the measures of decorrelation aimed to disguise the location of a base station were proposed in [75]. The proposed measures include hop-by-hop re-encryption of the packet to change its appearance, imposition of a uniform packet sending rate, removal of correlation between a packet's receipt and forwarding time and creation of multiple random areas of high communication activity. The multi-path routing and fake packet injection are used in this approach too.

9.5 Monitoring of Data Streams

When sensor networks are used for monitoring and surveillance, they should support distributed sensing of physical environment through measuring and aggregation of data in order to create the dynamic global view of the environment. These tasks generate various streams of measurement data within the networks. However, the data streams from sensors, sometimes referred as sequences of events, can be privacy-sensitive but have be used to monitor and detect events of interest.

One approach would be to use pattern matching on data streams from sensors. A privacy preserving pattern matching problem where patterns are defined as sequences of constraints on input data items means that sensor measurements will be evaluated as predicates privately, that is, sensors will not need to disclose the measurement values to the monitor.

Formally, a pattern P is given as a sequence $\{p_1, p_2, \ldots, p_n\}$ of predicates p_i defined on A. A privacy preserving pattern matching algorithm addresses the following problem: given a pattern P and an input sequence $t = e_1 e_2 \ldots$, find privately all positions in t where P matches t. The privacy preserving here means that sensor measurements will be evaluated as predicates $p_i(e_j)$ privately, that is, sensors will not need to disclose the measurement values e_j to the monitor. The framework and a solution of the problem based on SMCs was considered in [203].

9.6 Privacy Aware Routing

For security and safety applications, the reliability and trustworthiness of gathered data are critical. For example, querying temperature from a safety-critical facility must be performed in a trustful way. Commonly, data is routed through intermediate nodes where some of them may be compromised or data may be changed during transmission. At the same time, one would expect that temperature readings from sensors in the same area are correlated. When we aggregate that are expected to be correlated (data from sensors in the same area or data from the same sensor delivered via different routes), we would like to take into consideration the trustworthiness of these measurements as well (more trustworthy measurements should weigh more). Assuming that more trustworthy sensor nodes provide better protection of data they collect or transmit, a trust-aware routing can be seen as an approach for protecting transmitted data privacy.

A trust-aware routing assumes that all sensors in the network are assigned an initial trustworthiness level. Initially, trustworthiness of each sensor or groups of sensors can be determined from their context (e.g., properties of deployment area, sensor design, etc.). For example, deployment inside buildings may provide more physical protection to sensors compared to outdoor deployment, where intruders can have easier access to the sensor. In addition, we have to consider properties of the transmission channels between sensors. For example, data can be sent completely unprotected, integrity protected or encrypted with different degrees of encryption. All these options impose additional requirements on sensors, which sometimes are difficult to meet. For example, encryption will require extra computational resources in the sensor and will take time, energy, as well as key distribution arrangement. Generally, cryptographic protection can increase trust levels (and, consequently, privacy protection) but usually it carries additional cost and will reduce efficiency in terms of both time and energy consumption.

Consider a sensor query to collect movement detection measurements from inside of a monitored building. Privacy of these measurements may be critical since they reflect people activities inside of the building. Therefore, we need to make sure that the transmitted data are not exposed to adversary while routed via the network. Meanwhile, there are many different routes to deliver data from the sensor nodes. Some of them involve only sensors located inside the building. Other routes may be shorter and more time/energy efficient, but data should be routed through sensors located in a less protected outside area. The privacy protection can be achieved by choosing the route that is trustworthy enough to provided expected level of privacy. The required level of trustworthiness could still be achieved assuming that cryptographic protection is applied and the measurements are encrypted. However, in this case the time/energy efficiency would decrease, since both encryption and decryption are time and energy consuming. Moreover, in many cases such cryptographic protection is not available or provides only limited protection due to resource constraints of sensor nodes.

Thus, the notion of trust can be used to provide data privacy. The trust-aware approach privacy protection can be used when cryptographic protection is not available or is too resource demanding. The required level of privacy protection can be achieved by routing data via trusted sensors, even though such trusted routes are longer and may be more time/energy consuming. Given some trustworthiness requirements, we can select a route to transmit data from the sensors to more powerful intermediate nodes that protect data cryptographically, and then route protected data through lower trustworthiness sensors. The approach based on use of subjective logic [156] is proposed in [201].

9.7 Summary

Lack of privacy protection in WSNs is one of the biggest challenges threatening their successful deployment in the future. Not all privacy techniques developed for general network scenarios are appropriate for protecting privacy in WSNs due to the fact that many of the approaches introduce overhead which is too high for WSNs. Many privacy-related issues in WSNs can be solved by applying security mechanisms. However, some of them, for example, location privacy cannot be adequately addressed in this way. This is due to the ability of the adversary to observe and manipulate the packets sent over the WSNs. As a result, in the most cases privacy protection in WSNs is a trade-off between provided level of privacy and the overhead.

10

Radio Frequency Identification

In this chapter, we examine and outline security and privacy aspects of Radio Frequency Identification (RFID).

10.1 Brief Introduction to RFID

RFID is not a new invention. It has roots back to the *"Identification, Friend or Foe (IFF)"* systems used by the military since World War II for identifying aircrafts. The use we know today dates back to around 1970, but really took form during early 1980s. One early notable event was United States Patent No.3,752,960 dating back to August 14 1973 [246]. Here, the inventor described his "Electronic Identification & Recognition System" (see Fig.10.1). The same inventor, Charles Walton, also published another significant patent on "Portable Radio Frequency Emitting Identifier" [247] ten years later. Of course, what we recognize today as RFID has no single inventor, but these were significant early achievements.

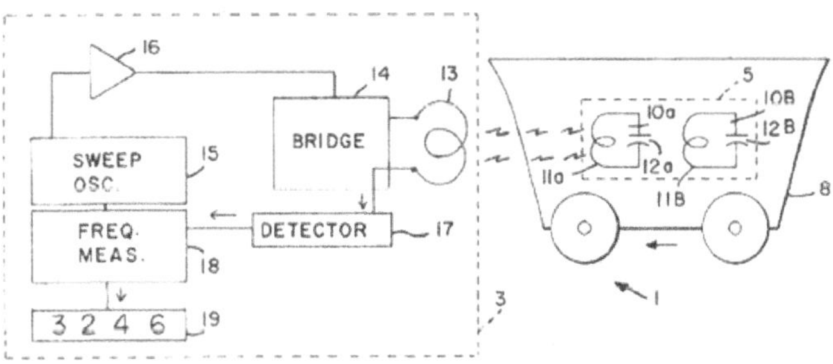

Figure 10.1 The "Electronic Identification & Recognition System"

RFID technologies are already deployed in large scale in, for instance, in electronic product code (EPC) settings. Use of RFID is on the increase and there seems to be a stronger drive towards more and cheaper devices rather than better and smarter devices (although both are increasing). A useful introductory text on RFID and privacy can be found in [171]. We also recommend the *IEEE Security & Privacy* magazine articles [118, 228].

10.1.1 What is RFID and what is it used for

An RFID device consists of a small and inexpensive microchip attached to an antenna. The device has very limited capabilities, both in terms of memory and processing capabilities. The chips can be very small and some RFID tags are tiny enough to be embedded in paper. Most RFID devices are only capable of transmitting a unique serial number up to a short distance and they do so in response to a request from a reading device. The RFID devices come in many flavors and can broadly be divided into two main types: Passive (low cost) RFID tags and Active RFID devices.

RFID is also being standardized and the International Organization for Standardization (ISO) has published several standards for RFID (including [141–145]). There are also industry consortia standards like the Electronic Product Code (EPC) standards. The EPC industrial standard are developed under the auspices of `GS1` consortium and is called the `EPCglobal` standard. The `EPCglobal` architecture is defined and described in [90]. The standard covers almost all aspects of EPC deployment, including safe disposal of used tags etc., but the security and privacy solutions are still a subject of criticism [130].

10.1.2 Passive RFID tags

Passive RFID tags have no internal power supply and rely solely on the electrical current induced in the antenna by the radio signal from the RFID reader. This will provide enough power to transmit a response. Since the passive tags do not need to include a battery, the devices can be very small. One example is the Hitachi μ-chip (2007). The μ-chip uses the frequency in the 2.45GHz area. It has a 128-bit memory for storing an ID number. The physical size of the chip is a minute 0.4m^2. The μ-chip is small enough to be embedded in paper. Since then even smaller units have been developed and the smallest units are now virtually invisible to the human eye under most conditions. Hitachi set a record for the smallest RFID chip when they produced the "RFID powder" chip at $0.05\text{mm} \times 0.05\text{mm}$ [135]. These dust-sized chips can still store a 128-bit number.

However, the size of the RFID chips is somewhat misleading as a size indicator for passive tags since the external antenna could easily be in the order of 100 times bigger than the chip itself. Figure 10.2 illustrates this. Alternatively, with a small antenna the RFID reader needs to be much closer to the tag, limiting the read distance down to the millimeter range. Inexpensive EPC RFID tags only cost around 5 cents (U.S) per tag, but specialized EPC tags can cost $\geq$100 times that amount. These specialized tags may be designed to be mounted on metal, which introduces difficulties for the antenna design, or for instance be designed to withstand gamma sterilization (food industry). There are also the so-called battery assisted tags. These hybrid devices are essentially passive devices, but with the benefit from an additional power source they can be more sophisticated and have more processing capabilities.

Figure 10.2 Wal-Mart passive RFID tag

The miniature tags, small and cheap by nature, can obviously be used in ways, whether intentional or not, that threatens our privacy. From a practical point of view, the most problematic aspect might be the pervasiveness of the passive tags. There are literally billions of these devices around and the security features are often very rudimentary, and one cannot expect much in terms of security and privacy protection except from naturally limited lifetime and limited range.

10.1.3 Active RFID Devices

In contrast to passive RFID tags, the active RFID devices have their own internal power source (battery) to generate the response signal. Active devices are, in general, more reliable (fewer errors) than passive tags, but they are also limited by the battery lifetime and are normally much more expensive than passive tags. The active devices can have a fair amount of memory and can perform substantial processing. Since the active devices have their own power source they can also transmit at higher power levels than passive tags. This allows the active RFID devices to operate at longer ranges, up to 100 meter is not uncommon, and to operate in "difficult" radio environments where passive tags would not have worked at all. Battery lifetimes between 5 and 10 years are not uncommon. The batteries are typically physically embedded (without any possibility to replace it) in the device and the lifetime of the device is decided by the lifetime of the battery. Active RFID devices come in many shapes and sizes, ranging from small and relatively inexpensive devices (coin sized, cost as little as 1–2 U.S dollars) to expensive (>$50) and relatively large devices.

Figure 10.3 Active RFID: The Autopass tag

Figure 10.3 depicts the AutoPASS tollroad device. Since the active RFID tags are relatively expensive and are used to track and protect valuable assets these devices can support cryptographic functions. Active RFID devices can therefore also afford to implement privacy enhanced access schemes and be able to protect captured and stored data.

10.1.4 Use of RFID

We have already mentioned a few areas where RFID is used. The following list should give an indication of the broad scope of possible uses for RFID solutions. The list certainly isn't exhaustive and the use of RFID is on the increase in more and more areas.

- *Inventory Control / Electronic Product code (EPC) uses*
  Replaces bar codes. Where bar codes would be limited to identifying the type of item, the EPC can additionally identify the specific item.

- *Food industry / Electronic Product code (EPC) uses*
  Tagging of food items, in combination with advanced logistics solutions, should allow for faster and safer delivery of food items from production to the customer. This is an important area since approximately 1.3 billion tons of food are lost or wasted annually (estimate for 2011 [128]). This is about one-third of the global food production and even a slight improvement here would be significant.
- *Electronic vehicle registration and passing of toll gates*
  RFID tags embedded in license plates are already in use and commercial solution exists. One example is the 3M Electronic Vehicle Registration System. Other systems, like the AutoPass system used in Norway (see Figure 10.3), offer efficient toll gate passing.
- *Parking permits*
  Many systems already exist for this purpose. Can be combined with vehicle registration systems and with ticketing systems.
- *Electronic ticketing*
  An example of this would be the Oyster card used on public transport services within the Greater London area of the United Kingdom.
- *Access control*
  Many systems exist which aims at providing "electronic keys" type of solutions. The key card access rights can be adjusted dynamically and one avoids inflexible solutions based on master keys etc.
- *Anti-theft protection*
  Shops and malls regularly deploy anti-theft RFID tags. These can be embedded into expensive garments or other expensive items and when passing through the shop entrance/exit the device will trigger an alarm unless the tag has been disabled by a shop clerk.
- *Transportation and logistics*
  This is a huge and diversified area. Suffice to say even very minor improvement in terms of reduced delays or reduced losses amounts to billions every year.
- *Identification*
  This ranges from livestock identification where cost control and disease control are important targets, via pet identification and to human identification. Human identification again ranges from embedding RFID tags into identify cards and passport to tags intended to be used for tracking individuals. This again ranges from tags used by soldiers, fire fighters and and police officers to tags intended to control criminal or tags used for safety purposed (tagging of senile citizens or tagging of children).

Needless to say, tagging of humans is a source of great controversy and it raises many deep ethical and moral issues.

10.1.5 The use of RFID for human identification

This is a subject which evokes strong emotions and it brings the privacy point home very strongly. However, emotional responses aren't always very useful and the subject has been surrounded by irrationality, fear and anger. This is not to say that human identification is unproblematic, far from it, but calm evaluation of the actual properties of the solutions is needed. Of course, one also needs to evaluate the existing solutions and proposed alternatives, before passing judgment.

One issue is whether or not ordinary citizens have any choice but to accept RFID for human identification. In cases where the government sanction RFID use it will become next to impossible to avoid it. In 2006 in the u.s., the Department of Homeland Security (DHS), produced a committee report called the "The Use of RFID for Human Identity Verification" [69]. The full committee advised caution in deployment of RFID for human identification purposes. However, interestingly enough, in a subcommittee input document the advice was not to use RFID for human identification at all [70]. The executive summary of the subcommittee input document summarizes the problems with RFID human identification [70]:

> There appear to be specific, narrowly defined situations in which RFID is appropriate for human identification. Miners or firefighters might be appropriately identified using RFID because speed of identification is at a premium in dangerous situations and the need to verify the connection between a card and bearer is low. But for other applications related to human beings, RFID appears to offer little benefit when compared to the consequences it brings for privacy and data integrity. Instead, it increases risks to personal privacy and security, with no commensurate benefit for performance or national security. Most difficult and troubling is the situation in which RFID is ostensibly used for tracking objects (medicine containers, for example), but can be in fact used for monitoring human behavior. These types of uses are still being explored and remain difficult to predict. For these reasons, we recommend that RFID be disfavored for identifying and tracking human beings. When DHS does choose to use RFID to identify and track individuals, we recommend the implementation of the specific security and privacy safeguards described herein.

In this book, we mainly look at the technical perspectives of privacy. In this respect, our main concern shall be technological concerns with RFID security and privacy. We will not further discuss the politics or privacy advocacy involved in RFID for human identification, but rather investigate whether the RFID solutions have the minimum necessary security features needed to support credible long-term privacy.

10.2 Basics of Security and Privacy for RFID

10.2.1 The "Kill Command"

Cheap RFID tags are not very capable devices. These tags don't have cryptographic processing or other advanced features. What modern passive RFID tags often have is a `kill command` that is used for destroying/disabling the device at checkout. The kill command is activated by a unique/individual (per tag) code sequence and it renders the tag unreadable [118]. There is no cryptographic protection or similar protecting the kill command, in some cases the scheme relies only on the assumption that the code sequence (32-bit code for EPC-type of tags) will not likely be triggered by accident. However, for the consumer it is almost impossible to know if an RFID tag has been killed. The problem with user control and assurance still remains. Another problem with the kill command is that even if one attempts to disable the chip by the kill command it is not too difficult to block or disrupt the radio signals. Thus, the "kill" may not succeed even when there are good intentions to carry out the kill command.

The kill command feature may also be subverted and used as an attack against the RFID tag. This may not be a privacy problem, but it illustrates that the kill command feature has several problems. This reduces the overall usefulness of the scheme.

There problems with the kill command have been recognized and there are proposals for how to solve the problems. One suggestion by IBM, captured in US patent no.`7,737,853 B2` [190], is to implement a means for a mechanical disabling feature. The authors note that "*...consumers are afraid that the kill command may not permanently 'destroy' a tag. The entity who made the tag may also have means to reactivate it.*" [190]. The scheme works by having a simple means to physically cut the wires between chip and the antenna. As stated in the patent, this will not guarantee that the device cannot be reactivated, but it will at least take an effort to do so.

10.2.2 Privacy problems with passive RFID tags

Inexpensive RFID tags are being used as replacement for barcodes. The EPC tags have two distinct characteristics that set them apart from traditional printed barcodes.

- *An RFID tag carries a unique identifier*
  The traditional barcode only indicates an object type. The book barcode depicted in Fig.10.4 identifies the book title etc., but not the individual book copy. A barcode printed on a box may state that it contains gouda cheeses and identity the manufacturer. An EPC tag, on the other hand, carry an extended serial number that not only identifies the type of item, but also the individual gouda piece. This permits fine-grained control over product distribution and permits tracking of individual inventory items, which for food items typically allow production dates and best-before information to be associated with the item.
- *An RFID tag will be read by radio contact*
  A barcode scanner must make close-range optical contact to read a barcode effectively. In contrast, an EPC tag may be read just by being placed in the vicinity of a reader.

ISBN 978-87-92329-32-5

9 788792 329325

Figure 10.4 Book barcode

The resolution of RFID identification is problematic seen from a personal privacy stand. If you buy an RFID tagged chocolate bar and carry it with you then you can also be tracked. The tracker may or may not know your identity, but they can certainly track you if they have the appropriate reading device. And with advanced data mining technologies it is not too difficult to establish the identity either, particularly if your tracker also happens to know that you like gouda cheese and chocolate bars. The above also serves to illustrate that collections of seemingly innocent information, such as your preferences in cheese, chocolate bars and shopping habits, might actually be used to identify and track you.

The risks of being tracked is mitigated by the limited range of the tags and the use of kill commands could eliminate and/or reduce the risk even further. However, use of the kill command is not universal and in many cases the kill command is not implemented. For identity cards or other devices that should have long lifetimes the use of a kill command is also only appropriate when the device is to be terminated/discarded.

10.2.3 Privacy problems with active RFID devices

The traditional active RFID devices are comparatively expensive and there are much fewer of them than the EPC tags. However, the active devices do have substantially longer range and are thus more exposed to illegitimate reading/eavesdropping. The active RFID devices are also much more capable devices. They can store a lot more (potentially privacy sensitive) data and do a fair amount of active processing. These devices would likely also contain some cryptographic processing capabilities. Another characteristic of the active device is that it is typically embedded in items that the users keep with them (associated with them) for prolonged periods of time. For instance, RFID-enabled identity cards would typically travel with the user. License plate tagging and the AutoPass toll road payment scheme are other examples. Yet another example would be automated teller machines (ATM) that uses RFID-enabled ATM cards. There is, therefore, every possibility that we will carry a number of active RFID devices with us more or less continually. If these devices leak information to the surrounding environment then we are susceptible to being identified and tracked.

10.2.4 Classification of RFID Attacks

Given that security is the foundation for credible privacy it makes sense to investigate the security of RFID devices and the security risks that RFID devices face. In [189], the authors have outlined a classification scheme for RFID attacks.

The attack vectors include:

- *Permanent disabling of tags (by physical means)* This includes tag removal, tag destruction, abuse of the kill command. The previously mentioned IBM patent for disabling RFID tags to improve privacy may be subverted and used in a "tag destruction" attack.
- *Temporary disabling of tags (by physical means)*
  This include use of a Faraday cage (aluminium foil surrounding the tag) and use of radio interference to disrupt communications. This could even include devices devised to enhance privacy, like the "blocking reader" described in [193].

- *Removal or disabling of RFID readers (by physical means)*
  RFID reader can be stolen, destroyed or otherwise disabled.
- *Relay attacks*
  This would be a man-in-the-middle type of attack. The adversary may intercept, modify and selectively forward message between the reader and the tag.
- *Tag attacks*
  Cloning of tags and spoofing of tags. The cloning is simply that of copying a tag and making an identical clone. Spoofing is somewhat more sophisticated in that one then uses a more capable device to emulate the tag. A spoofing device would potentially work against a whole range of devices and not just one particular type.
- *Logical reader attacks* In many cases RFID communication is not authenticated or it is only weakly authenticated. Impersonation (masquerade) may therefore easy or at least feasible in many cases. Eavesdropping is also a considerable risk when the security is weak.
- *Network protocol attack*
  The tag number is really only a number and as such it does not convey too much information. However, the tag number is logically associated in the database with an item. By attacking the protocol one may for instance fool the checkout procedure into believing that the item to be purchased is a cheap inexpensive item, while it in reality is quite expensive. This may represent a real risk in do-it-yourself unattended shop checkout systems that are being deployed in several countries.

The authors lists several other attacks, but the above attack list should suffice to highlighting vulnerabilities and risks in RFID security. In [181], the authors are specific and investigate security weaknesses in passport cards, driver licenses with tags and similar identification cards. They also propose defenses and mitigation strategies. We shall return to this in the subsequent section (Sec.10.3).

10.2.5 Classification of RFID Threats to Privacy

In [118], the authors outline a set of RFID related threats to personal privacy. The main threats are:

- *Action threat*
  This is a threat that is triggered if many tags are moved simultaneously in a shop. The fact that many tags are moved may indicate a case of

shoplifting etc. This in turn may trigger (video) surveillance and direct action by the shop. The problem is that there may be many legitimate actions that can cause multiple tags to be moved simultaneously.

- *Association threat*
  The customer identity can be associated with the RFID tag. Loyalty cards etc. are in use in many shops. Other EPC tag can then be linked with the loyalty card and a database of customer spending patterns formed. The threat is similar to a bar code association threat, but since the EPC codes have higher resolution (identifying specific instances instead of type of object) the privacy threats worsened.
- *Location- and identity threats*
  Pervasive deployment of RFID technology means that it will be quite easy to place covert RFID tags such that one can track users. The tracking may be of unidentified users (called a constellation threat in [118]), but one may also identify the users and track specific users. It is noted that using multiple RFID sources may mean that a resourceful adversary can correlate information and gradually establish a user identity. The user will then lose his/her identity- and location privacy.
- *Preference threats*
  Obviously, the RFID technology can be used to establish a pattern of the user behavior. This includes learning user preferences. This extends beyond knowing what products you buy since it is also possible to identify each product individually. Thus, one can build up a history of your preferences which includes knowledge on when and where you bought certain items and conceivably even the price you paid for the products.
- *Transaction threats*
  If your whereabouts can be traced then one may also infer your activities. That is, one can derive information about who you meet and what you do.
- *Breadcrumb threat*
  The breadcrumb threat arises from an association threat. If you are associated with a certain set of RFID tags/devices, then the signature of set of RFID tags/devices de facto constitutes an identity. Thus, if somebody gets access to the emergent "associated identity" then they can potentially impersonate you. The threat arises from wrong use of data, but this does happen quite regularly and the threat is therefore quite serious.

Note that many of the above "threats" may also be seen as features. Certainly, for the issuer of a loyalty card the ability to build a personal consumer preferences database may be seen as a feature. The customer may indeed also see this as a feature, and the only threat would then be if external parties get unsolicited access to the information.

10.3 Advanced Security and Enhanced Privacy for RFID

10.3.1 Theoretical Frameworks and Modeling of Privacy for RFID

In order to get the security and privacy one wants or needs it is essential to be clear on what the problem is. To this end, one needs a model of the RFID context, the entities, the communications channels, the adversary/intruder and the privacy and security goals one has. The model, as all good models should be, is a simplification of the real-world, but should still capture the essence of the problem at hand. The models should make the problem at hand transparent and tractable.

The "On Privacy Models for RFID" paper from 2007 by Vaudenay [245] is one effort in providing a formal model for security and privacy for RFID systems. The model has a powerful intruder who can monitor all communications, who can trace tags within a limited period of time, corrupt tags, and get access to side channel information on the reader output. The model additionally caters for intruders that do not have access to side channels, and these intruders are called "narrow adversaries". Furthermore, intruders are classified according to their ability to corrupt tags. They have properties denoted "...strong, destructive, forward, or weak...". In the model, Vaudenay operates with eight privacy classes and investigates them to find out what is possible with respect to privacy. Somewhat depressingly, Vaudenay reaches the conclusion that strong privacy is impossible when facing a powerful intruder. Privacy can be attained for other circumstances, depending on security schemes (defined in terms of cryptographic properties) available to the defender. Unsurprisingly, Vaudenay proves that public-key cryptography (within the model) can assure the best privacy solutions. Sadly, this is also the most costly solution in terms of memory requirements, power requirements, processing power and bandwidth usage. Cheap EPC type of tags simply cannot support public-key cryptographic primitives.

Vaudenay's model attracted quite a bit of attention in the privacy research community and soon new and improved models appeared. In [194], the authors analyzed the Vaudenay model and found that they could simplify the eight privacy classes down to just three classes. Furthermore, they declared that strong privacy was achievable. The result was claimed for the given (revised) model, but the authors also claimed that it should be achievable in practice. The authors also question the realism of the strong adversary. They do so since they believe that the added cost of protecting against a strong adversary would in practice be impossible to defend. That is, cheap tags cannot afford to include public-key cryptographic primitives and a reliable randomness source. To argue their case, they point out that "Due to the short communication range and infrequent access properties of RFID tags, we believe it is not necessary to assume the presence of powerful adversaries." [194].

Another paper on the same subject is [74]. The paper has two main results, the first being that the Vaudenay model is inadequate in the sense that privacy cannot be attained under Denial-of-Service (DoS) attacks. DoS attacks are easy to carry out so this is an important result. Furthermore, the second result is a claim to be able to afford so-called semi-destructive privacy using only symmetric-key cryptography. Again, we should stress that the results are relative to the given model, but symmetric-key cryptography is a lot more affordable than public-key cryptography so the result is indeed important.

The paper is also important in the way the authors emphasize efficiency and suitability for low-cost tags. In [186], the authors demonstrate that "a pseudorandom function family is the minimal requirement on an RFID tags computational power for enforcing strong RFID system privacy". They have also devised an RFID protocol which satisfies these requirements, and which in theory should outperform existing solutions. Again we are mindful about the assumptions made, but it is yet another interesting and promising result.

The point we have made about the premises for the model should not be taken lightly. In [132], the authors challenge what they describe as "insufficient generality and unrealistic assumptions regarding the adversaries ability to corrupt tags". In [76], the authors challenge the notion of strong privacy used in other models, claiming that they are not as strong as they could be.

Ultimately, one should realize that *efficiency* is extremely important to low cost tags, and that most of the above models don't really capture this. The tags must be extremely cheap to produce, cost very little to deploy and must run off virtually no energy at all. Operating cost must be contained and most security management solutions would simply be too expensive. All in all, this invalidates many of the assumptions made and we are still on the outlook for realistic models for low-cost devices.

10.3.2 Physically Unclonable Functions

First, it should be noted that ultimately there may not really be such a thing as a *physically unclonable function* (PUF). In the same vein as one refers to smart card technology as being *tamper resistant* and not *tamper proof*, one ought to have called it a cloning resistant function However, the name physically unclonable function and the acronym PUF is widely in use, and so we shall use it too.

Tamper resistance and resistance against cloning are obviously desirable attributes for an RFID tag to have. In the paper "Enhancing RFID Security and Privacy by Physically Unclonable Functions" [220], the authors present a system model and highlight how PUFs can be used to improve the security and privacy for RFID tags.

The most novel part of the paper is the part where one use PUFs as a tamper-resistant authentication key storage. The key is not actually stored on the tag, but rather it is extracted and reconstructed from the physical characteristics of the tag. The idea is that any attempt to tamper with the tag will also change the authentication key, and thereby effectively making the key invalid. Thus, the tampering will in effect make the tag unusable.

A PUF consists of special functions P, which is integrated into the tag T. The function P is unclonable and it is a so-called *noisy* function. The unclonability stems from the randomness in the noisy characteristics, something which is introduced during manufacture. That is, the manufacturing process of the noisy function P will inevitably lead to unpredictable and to all appearances random characteristics being embedded in the function. The characteristic will typically be extracted from the specific statistical properties of gate delays and similar. The characteristic should also, with overwhelming probability, be unique for the particular tag [220].

The PUF is then used in a challenge-response authentication scheme. The challenge c, called a stimulus signal, is then fed into P and the output is the response r'.

$$P(c) \rightarrow r' \tag{10.1}$$

The output of P to stimulus c is then unique to the specific tag T. Function P is not a "clean" function in the sense that the output to some extent will be determined by environmental changes to the physical circuit that realizes P. This could be due to differences in current voltage levels, strength of induced stimulus signal, ambient temperature etc. So, within bounds, the output r' is not entirely fixed. To sort out this inconvenience, the PUF also includes a so-called fuzzy extractor. The fuzzy extractor in tag T, which knows about the natural variances of r for P, will be able to normalize the output and map r' to r. There are boundaries to what the fuzzy extractor can tolerate, but the idea is that when applying the fuzzy extractor the output will be fixed.

$$FuzzyExtractor(P(c)) \to r$$

One problem with the PUF approach is that the basic function is static. That is, for a fixed (T, P, c)-tuple, the output will always, with the help of the fuzzy extractor, be the same r. This is problematic if P is used naively in that an adversary listening to a (c, r) exchange will be able to replay the sequence later on. This problem can be remedied, but it takes effort and introduces additional complexity to the authentication scheme. Another issue is that the manufacturer of the tag will not know the "key" that the tag uses. So an RFID system using a PUF enabled tag must learn and record a set of (c, r) pairs by challenging the tag before deploying it.

Given that the key is embedded in the PUF, it is not possible to verify the output of P by any external means except from P itself. This makes it harder when it comes to system design and it makes it more difficult to design an effective and efficient security and privacy architecture for the RFID system. The nature of the PUF construction also requires a lot of statistical testing and verification. It is after all essential that the function is not clonable, and an easily predicable function would be clonable as well as a function with a narrow output (.i.e. with too few unpredictable bits in the output). So, while the PUF concept clearly is an interesting proposal for solving RFID security problems, it is not yet a mature solution ready for deployment.

10.3.3 RFID Services and Associated Privacy Solutions

In the paper "A survey of RFID privacy approaches" [183], the author attempts to given an overview of RFID privacy approaches. The paper starts out by observing that there is a huge and growing body of RFID privacy research papers made available, yet no real or realistic solution seems to emerge. Part of the problem is that RFID tags are many things and they are used for a rather large and diverged set of applications and services. When devising detection and protection schemes for security and privacy one should always start by identifying what it is that one wants to protect. That is, what are the assets? Then one should attempt to reason about what are the risks and threats with regard to those assets. Are there vulnerabilities? Are there any specific type adversaries one foresees? And of course, one should balance this out by trying to assess the worth of the assets and the cost of protection.

So, given the diversified set applications and use areas one has for RFID deployment, the search for a single silver-bullet solution seems deemed to fail before it even starts. But, this is not to say that one should abandon all hope of reasonable and credible protection.

In [183], the author defines four different services that RFID offers. It would then be useful to see what can be done to protect theses services, keeping in mind that even a partial solution could be useful in some market segments.
The services are:

- *Identification (of items).*
  This could range from a complete identifier to a single binary sold/not-sold or paid/not-paid tagging. EPC tags have standard formats for representing this type of information.
- *Alerting*
  Alerting functions include the anti-theft usage typically found in shops.
- *Monitoring*
  RFID tags are used extensively for logistics purposes and monitoring is a primary RFID services.
- *Authentication*
  For expensive items, one cannot rely on unsupported claims made about identity. Authentication is then an essential protection against subversion of the other three services.

None of the above functions automatically caters for privacy. Many, if not most, RFID systems are designed to support identification and traceability of the tags. So, even assuming credible authentication, which may solve many security problems, one is still left without credible privacy solutions.

10.3.3.1 Identification

Unauthenticated identity claims can easily be forged. Cheap RFID tags can also be cloned and otherwise physically manipulated.

However, while attacks against the tags certainly are feasible, they may not be practical. That is, when one considers the economy of attacking RFID systems, it should be readily apparent that attacks against tags in a one-at-a-time style simply do not scale very well while the cost of attacking will. However, attacks against the *RFID reader(s)* may scale effortlessly. These attacks may be physical or they may be logical (attacking reader software or reader connections). Of course, in an RFID system one can attack the supporting system too (see Section 10.2.4).

All of the above attacks on the identification can more-or-less automatically become attacks or tools for attacks on privacy. That is, we don't so much care about identification of a tag per se, but if the tag can be associated with a person one can soon enough have a privacy attack.

One suggested method of protection against illicit identification is the use of radio blocking equipment like the earlier mentioned blocking reader [193]. Kill commands and physical destruction of tags may also work, but one must then be able to locate the tags. For the kill command, one must also possess the PIN code used for the particular tag. Faraday cages work well in theory, but are cumbersome and impractical in use.

10.3.3.2 Alerting and Monitoring

Alerting and monitoring depends on the identification service. Essentially, they therefore suffer from the same problems that does identification. That is, from a privacy point of view, alerting and monitoring are obviously a greater danger. The effectiveness of the attacks would here strongly favor system attacks. Logistics systems, by their very design, provides automated alerting, monitoring and tracking features. A successful system attack would, therefore, be very effective.

10.3.4 Authentication in RFID systems

In Section 10.3.1, we more-or-less declared that strong security is not possible for cheap tags. This subsection on authentication in RFID systems may thus seem vain. But, we still have active devices, which can afford cryptographic support, and we have minimal solutions that aim more at increasing the cost of an attack than truly to prevent them. One may wonder what use it is to "increase the cost" of an attack, but in a system perspective this makes perfect sense since it is a defense against scalability of attacks. The overall system should easily survive attacks that only targets one tag at a time.

In basic RFID protocols, the tags identify themselves to the reader. Slightly more advanced protocols will also include authentication of the tag identity to the reader. Unilateral authentication is valuable, but it has inherent limitations. After all, readers are relatively cheap and so a reader masquerade attack is easy to carry out.

In "Mutual Authentication in RFID" [207], the authors proposed several schemes with mutual authentication. That is, schemes in which the tag also authenticates the "system". The "system" is physically represented by the RFID reader, but if one assumes that the reader is online during authentication then the reader need not store any security credentials and take any security decisions. The proposals in [207] make different assumptions on computational power of the tag and the strength of the authentication schemes are relative to this. The simplest scheme is based on the use of a pseudo-random function and a predefined tag state (key). The tag is required to hold intermediate values, but the exchanged data does not directly include the tag identity. In the Vaudenay model, this amounts to narrow-weak privacy [245]. The other and stronger authentication protocols depend on more processing power and more memory at the tag, and in the extreme this amounts to use of public-key cryptography.

There exists several papers extending or complementing the properties of the protocols in [207]. In [195], the authors came up with a classification scheme for RFID authentication protocols, and they show how certain properties cannot be attained for synchronized protocols. In [31], the authors investigate the properties of the mutual authentication schemes in cases where the tags have been corrupted.

10.4 Summary

This chapter on RFID has mostly been a survey over the RFID landscape. The emphasis has been on security and privacy aspects and we have seen that many of the RFID uses easily lead to problems.

Section 1 introduced the RFID types and use areas. Our investigation has mainly been focused on passive RFID tags. These have limited memory, no power except from radio induced power and very limited processing capabilities. This has a strong impact on the security and privacy features one can afford to support on the tags. Active RFID tags are potentially relatively powerful, and the more advanced model would quite effortlessly be able to support mutual authentication and even encrypted communications. One might here support challenge-response protocol akin to the UMTS AKA protocol (see Chapter 7) or even public-key protocols. We briefly highlighted some of the many different usage areas and lesson learned should be that RFID use is pervasive and ubiquitous. So much so, that tagging of humans, directly and indirectly, is also done quite routinely, albeit not in the general public.

In Section 2, we started to investigate basic security and privacy options as well as establish what the problems are. We first investigated the `kill` `command` and associated methods. While useful, the kill command clearly does not solve our RFID security and privacy problems. The kill command, as an answer, also illustrated the importance of getting the questions right. To that end, we started to investigate the privacy problems with RFID and we followed up on this by classifying attacks on RFID and by classifying privacy threats.

In Section 3, we informally investigated formal models for security and privacy and more security and privacy solutions. In typical cryptographic papers, there is often a strong emphasis on the theoretical models. The system is modeled in some formalism and the various entities and system properties are encoded in the said formalism. One can formulate research questions set out to prove of falsify claims within the model. Several models exists, but there is now convergence on models based upon the Vaudenay model [245]. We briefly investigated use of physically unclonable functions (PUFs) and looked more into the basic RFID services (`identification`, `altering`, `monitoring`, `authentication`) and their associated problems with respect to privacy.

In summary, we may conclude that RFID potentially poses huge threats to our privacy. The short range of passive tags provides some relief since it puts an upper limit to the efficiency of an eavesdropping attack. Not that such attack need be very difficult though. However, that be as it may, the pervasiveness of the tags is perhaps what should cause the most anxiety and, in particular, when combined with mandatory (in practice) for use system including identification schemes (passport, drivers licences etc) and always-with-us items such as credit cards and mobile phones. We note finally that from a systems perspective the largest threat to our privacy may ultimately not be tracking and tracing of single tags, but rather collective tracking and tracing of collections of tags. In this perspective, we should be more worried about security and privacy at the central RFID databases than about intrusions and attacks against the RFID radio interface.

11

Privacy and Trust for the Internet-of-Things

In this chapter, we examine and outline security and privacy regarding the Internet-of-Things (IoT). The chapter is inspired by the "Reflections on Trust in Devices" paper [176] (by one of us). This is particularly apparent in the focus on trust, which is a very important aspect for IoT devices and services. Trust and trustworthiness are higher layer concepts, and in this context security is a supporting function for trustworthiness. Credible privacy may be seen as a prerequisite for consumer trust, but it is also an end to itself.

11.1 Assumptions About the IoT Context

In the last chapter, we investigated RFID and it may seem that there is an overlap between IoT and RFID. And indeed, there is an overlap for active RFID devices as these could very well support IP connectivity. To clarify, in this chapter, we choose to classify IP-enabled active RFID devices as IoT devices.

11.1.1 Devices and Services

The IoT environment is a highly diversified and heterogeneous environment. There will be a multitude of different devices and there will be a huge number of services delivered, at least in part, by IoT devices. Some of the devices will be personal devices like our mobile phones while others will be anonymous, ubiquitous, almost undetectable and ephemeral. They will range from relatively expensive and high-assurance to cheap and unreliable; some will enjoy substantial physical protection while others will be distributed into hostile environments seemingly without any protection at all.

We shall henceforth choose not to classify personal devices, such as the mobile phone, as an IoT device, although these personal devices will undoubtedly be "on the internet". And of course, internet connectivity is a tautological assumption for IoT devices. What the IoT devices all have in common is the ability to communicate, the ability to compute results and being "unattended". Some will be highly dedicated and others will be more general purpose, but at the end of the day they should all deliver or perform some kind of service.

ASSUMPTION 1 (COMMUNICATION) – *The IoT device has IP connectivity.*
The IoT device need not be always-on and the device may not always get connected when it needs to, but in general it will be connected when needed.

ASSUMPTION 2 (CONDITION) – *The IoT device is unattended.*
The IoT device will generally be an unattended device. From a security point of view "being unattended" is therefore a premise.

11.1.2 Further Assumptions

In our discussion, we will examine Human-to-Machine interactions and the related security and privacy aspects. Many of the observations may apply to Machine-to-Machine (M2M) as well, but privacy aspects are not normally a central part of M2M communications, although M2M networks, may for instance, handle privacy sensitive data.

ASSUMPTION 3 (HUMAN-TO-IOT) – *Human-to-Machine aspects only.*
We choose to restrict our investigation to human-to-machine aspects.

As a generalization, we assume that device access is wireless, i.e. that the communication is not directly observable or verifiable to a human consumer. The actual communications technology does not matter too much; the important point is that the human user lacks a reasonable, effective and efficient way of determining exactly which device he/she communicates with. Some systems will be designed explicitly to provide this type of assurance, but we cannot automatically or generally assume it to be the case.

ASSUMPTION 4 (INTERFACE) – *The user cannot verify the connection.*
Assurance through verifying the physical connection is, in general, not possible. The user cannot, therefore, make assumptions regarding which device he/she is connected to without explicit mechanisms to provide this assurance.

Another assumption made is that the human user does not know the exact referential identity or system address of the IoT device. For example, a human user may know the street address at the ATM being used, but its highly unlikely that he/she knows the IP-address or the device number of the ATM. We therefore have that the human consumer in general cannot be 100% certain with which device he/she is interacting and furthermore that she/he does not know the identity and/or address of the device. We may assume that the human user is aware of the service being requested and we may also assume that the human user has indication about what type of device he/she is interacting with (or intended to be interacting with).

ASSUMPTION 5 (IDENTITY) – *The user does not know the device identity.*
The human users may believe they know the identity or address, but this cannot be relied upon (see Assumption 4).

The ATM example above illustrates the ATM interaction through a human-to-machine interface. The ATM may well have IP connectivity, but the human-to-machine connection is clearly not through IP connectivity. To avoid confusion, we therefore assume that the human is communicating through a proxy device. This proxy device may be the mobile phone or another personal device.

ASSUMPTION 6 (HUMAN PROXY) – *The human user will use a proxy device.*
The general assumption is that the human user will use a proxy device to interact with IoT devices.

ASSUMPTION 7 (TRUSTED PROXY) – *The proxy device is a trusted device.*
We axiomatically define the human proxy device to be a trusted device. This does not mean that the device is trustworthy, simply that the human user has chosen to trust the device.

The two last assumption are specific to our purpose. Together with Assumption 3, this defines a subset of the full IoT universe.

11.2 Trust and Trustworthiness

Given the above assumptions, it should be evident that the human user will need assurance when connecting to an IoT device. The assurance is a little different from normal authentication in that the human user cannot be assumed to know the identity of the device. Therefore, corroboration of device identity may not be meaningful to the human user. On the other hand, the user can be assumed to know the type of service requested and he/she may know the identity of the service provider. That is, when purchasing a sandwich from a vending machine the user knows about the "service" and he/she knows about the service provider (perhaps only by recognition of vending machine type or the brand name). From a human perspective, the important question is "Can I trust this device/service?".

11.2.1 What is Trust?

Trust is a complex, fine-grained and multifaceted concept. A thesaurus definition of trust will typically highlight reliance on the integrity, ability or character of an entity, a person, an institution or an organization. Trust can be further explained in terms of confidence in the truth or worth of an entity. The challenge in our context is that we need to assess human "confidence in the truth or worth" of non-human IoT devices.

We have that trust in one area does not imply trust in another area, although human trust has a tendency to be influenced this way. For example, you may trust Google to be your search provider on the internet, but this should not by itself automatically lead to trust in other areas. However, Google has multiple services and implied trust is probably involved in our trust decisions for related (to the company) services. Thus, a branded device may be trusted, at least partially, since the user associates brand traits with the branded device. We observe that this type of trust is not grounded in the device itself.

Another aspect of trust is that it normally is conditional and confined. You may trust an acquaintance and loan him/her a $100 without further ado, but you probably will not do the same for a $10,000 dollars. With respect to conditional trust, we note that one may be inclined to trust a colleague with company information since we share the "company" context. Outside the "company" context, you may find less reasons to trust the same colleague.

Trust is also about choice, dependency and willingness to take risks. One often has no choice but to trust someone/something. An infant must trust his/her parents and an individual must (willingly or not) trust the government to a certain degree to get by in society. Little real choice and strong dependencies highlight the fact that there may be pronounced asymmetries in the trust relationships. Obviously, knowledge and available information also plays a role. A history of betrayal clearly does not instil trust. Colloquially, an individual known to keep her/his word at high cost will appear as someone that can be trusted. The level of risk also enters into the equation. To trust a vending machine with a couple of coins to gain a soft drink may be an easy choice. The machine may cheat, but the maximum losses are not high and realistically the chances of the machine not dispensing the soft drink are fairly low. The book "Liars & Outliers; Enabling the Trust That Society Needs to Thrive" [225] is dedicated to trust and trust concepts, and it provides a lot of background and analysis of trust. The book is recommended reading to gain a fuller picture of trust in a security-related context.

11.2.2 What is Trustworthiness?

This concept is about beliefs in *ability* and *intention*. Is the corresponding party able to fulfil its obligations? And to which extent is the corresponding part interested in fulfilling its obligations?

For humans there is an ethical dimension to being trustworthy; this notion clearly does not apply to devices. Instead, for devices, one is left with a probabilistic notion of intention and ability (i.e. the probability that the device will "intend" do as expected combined with its ability to actually do so). The "ability" to behave as expected must be present even under adverse conditions. Resilient design, proper security and responsible management is needed for supporting the "ability" aspects.

11.2.3 Perceived Trustworthiness and Trust Backlash

It may be hard to assess trustworthiness. So much so that parties may optimize the *perceived trustworthiness* rather than improving the actual trustworthiness. For instance, it may be cheaper to provide a well-known security solution that doesn't fit the system or the problem rather than to re-design the system to provide credible security. Or even, to promote outdated security solutions or security solutions that do not address the problem at hand. The strategy of promoting ineffectual security, often termed "Security Theater" [223], may ultimately lower the trust. That is, claimed trustworthiness that fails may lead to a trust backlash. This, of course, presumes that the affected party is able to detect the breach and that he/she actually attributes it to the offending party.

11.3 Trust in Devices

An IoT device consists of physical hardware (processor, memory, I/O hardware, sensors/actuators etc), software (firmware, operating system, drivers, applications) and a power source. The overall trustworthiness of the IoT device depends on these components themselves being reliable and trustworthy. The device and its components must not only act as expected, but they must be able to do so in a hostile environment. In this section, we examine trust and trustworthiness with respect to software and hardware components. We also briefly examine the mobile device, which we have assumed to be the trusted human proxy device.

11.3.1 Trust in Software

11.3.1.1 Can Software Be Trusted?

Ken Thompson is co-inventor of Unix and the inventor of the `B` programming language, which preceded the `C` language. He is also co-inventor of the recent `Go` language. In his 1984 ACM Turing Award acceptance speech, Thompson presented a though provoking paper entitled "Reflections on Trusting Trust" [237]. There Thompson outlines an attack in which an intruder has access to the source code of a `C` compiler and then modifies the binary `C` compiler to contain Trojan code. The attack code is specifically targeted towards compilation of the `login` program, and the effect is that `login` will accept either the intended password or a predefined password chosen by the intruder. This is certainly an interesting attack, but an inspection of the source code will immediately reveal the attack code. So Thompson outlined a second piece of Trojan code placed in the `C` compiler source code. This code affects how the `C` compiler compiles itself from source code. That is, the infected binary compiler (the Trojan) will always ensure that re-compilation of the compiler will result in the Trojan code being included in the newly compiled compiler. We denote the clean source code `cc.c`, the clean binary compiler `cc`, the infect source code `Trojan.c` and the infected binary `Trojan`. The procedure is as follows:

1. *Start state*
   Start out with `cc.c` and `cc`. Create `Trojan.c`, which contains the `login` and `cc` subversion code.
2. *Compile the Trojan*
   Use `cc` to compile `Trojan.c`. Outcome: `Trojan`.
3. *Clean-up and Subversion of `login`*
   Remove `cc` and `Trojan.c`. Rename `Trojan` to `cc`. Keep `cc.c`.
   Use the infected `cc` to compile `login.c`, thus generating the infected `login` program.

The end result assures that source code inspection will never reveal the Trojan code. According to Thompson, the lesson to be learnt is [237]:

> The moral is obvious. You can't trust code that you did not totally create yourself. (Especially code from companies that employ people like me.) No amount of source-level verification or scrutiny will protect you from using untrusted code.

This does not imply that no software can be trusted, and by and large we tend to trust software to do as intended. It is furthermore clearly impractical to "totally create" all software we need. The trust we have in software is facilitated by positive reputation and by association with a trusted vendor. But proof of trustworthiness may be hard to come by.

11.3.2 Trust in Hardware

That software cannot always be trusted and that software may contain malware is no big surprise. One approach to strengthen the trustworthiness is to have dedicated hardware to enforce security and privacy properties. This type of hardware exists and many software security issues may be solved or mitigated by using it. However, use of hardware is no silver bullet solution for security and privacy problems. There are several reasons for this and one is that hardware itself is not immune to attacks.

11.3.2.1 Hardware and Firmware

The processing unit(s) in a typical IoT device will contain millions of transistors. To trust the processing units, one would have to be assured that none of the circuits contain malicious instructions. In the paper "Hardware Trojan: Threats and Emerging Solutions" [56], the authors outline the design of a generic hardware Trojan. They point out the risks with respect to high assurance devices and for trust in safety-critical applications. Use of automated design tools is one way of inserting malicious instructions, analogous to the Trojan `C` compiler that subverted the `login` program. New hardware designs are almost always built with automated high-level design tools and around pre-made building blocks with automated wirings etc in place. Thus, unless one is fully assured of all these tools (software and hardware) and the fabrication process itself, we have that (to paraphrase Thompson [237]) "You can't trust hardware that you did not totally create yourself".

There is a growing body of papers on this topic. In [236], the authors present a classification of hardware Trojans. They also discuss various ways of implementing Trojans. In [152], the authors report their experiences in practical design and implementation of hardware Trojans. The approach was to demonstrate feasibility and not to make any realistic Trojans. The Stuxnet worm [106] is an example of hybrid malware containing "traditional" software components and additionally containing firmware payloads etc. To round off, hardware-based malware is feasible or becoming feasible. Firmware-based attacks are already with us. This imposes an upper limit to the trust we may have in hardware and firmware.

11.3.2.2 Trusted Execution Environments

One trusted execution environment concept is the so-called Trusted Platform Module (TPM). The TPM concept is standardized and it details a secure crypto-processor, which has secure storage and provides secure processing. Several implementations of the TPM standard exists, and are commonly referred to as TPM chips or TPM secure devices. The smart cards used in mobile devices in the 3GPP cellular systems (SIM card/UICC card) is another example of a trusted execution platforms [9]. These devices can be quite powerful and nowadays have a fair number of form factors. Figure 6.1 depicts a "normal" size UICC smart card. The 3GPP has also defined a trusted environment (TrE) to be used in 3GPP-based femto-cells [15].

11.4 Trust in the Mobile Phone

We now look at what possibly is our most personal and likely also our most trusted device, the mobile phone. In Assumption 6, we stated that the mobile phone will be a proxy device for interaction with IoT devices. But is there a basis for trust in the mobile phone? Is the security and privacy trustworthy?

11.4.1 The Smartphone as a Proxy

Smartphones dominate in mature mobile markets and it is the de facto baseline platform. A smartphone features a reasonably powerful computing platform (32-bit CPU, often multi-core), it runs a comparatively advanced operating system and it has plenty of memory. It also will be running quite a few third party applications.

Needless to say, the smartphone will support seamless cellular connectivity and will regularly support WLAN, Bluetooth and USB connectivity amongst others. The smartphone shares quite a few characteristics with a personal computer and will generally suffer from the same security problems as does a generic computer. With respect to privacy, the smartphone may or may not be more vulnerable than other computers per se, but given that smartphones are highly personal devices they are likely to be (or become) important targets for privacy intrusions.

11.4.1.1 Communications Security and Privacy

Cellular system, which includes GSM/GPRS, UMTS and LTE/LTE-Advanced, basically provide entity authentication and (over-the-air) link layer security [167, 173]. Notably, these systems do not provide end-to-end security. This is not so much that it is inherently impossible, but rather that they, for regulatory reason, cannot do so without also providing Lawful Interception capabilities. Since that would pretty much defy the purpose, these systems generally do not offer end-to-end protection. Chapter 7 provides an overview of communications security for the 3GPP systems. With respect to privacy, we note that identity privacy and location privacy are generally not fully supported (see Chapter 7). The 3GPP2 systems (cdmaOne/CDMA2000) offer similar protection to that of the 3GPP systems [218].

11.4.1.2 Trust in the Smartphone Hardware and Software

As with any hardware device the mobile phone may potentially be compromised by malware. The mobile phone hardware can likely be trusted, but one should bear in mind that the hardware will contain test circuitry, digital rights management circuitry etc. One may also find that the operator installs measurement software etc. unbeknown to the user (Ref. the `Carrier IQ` software [125]). These aren't malware per se, but they may nevertheless be used against the user. There is no reason to assume that smartphone software is any worse or better than software for other general purpose computing platforms. However, general purpose platforms have a large set of weaknesses and vulnerabilities and unfortunately there is little reason to suggest that the smartphone will fare any better.

11.4.1.3 Trust in the Secure Modules (Smart Cards)

The SIM card or the more recent UICC [9] represents a trusted environment for the mobile phone. The standardized functionality, which includes the USIM software component, is only designed to provide security functions for the cellular link [173], but the UICC may support additional applications to facilitate banking and eCommerce. Despite this, there is some inherent design restrictions that impede trust in UICC-based functionality. For instance, all user I/O to/from the UICC is by means of the mobile phone and there is also generally no authentication between the SIM/UICC and the mobile phone. The UICC/USIM may be a trustworthy basis for cellular link protection, but its not a silver bullet solution for security in the overall smartphone.

11.5 What About Privacy for IoT Access?

In our assumptions for IoT devices we had that these are not personal devices. We furthermore had that they, as a generalization, communicated through wireless communication. When it comes to mobility for the IoT devices we do not have any assumptions, but the IoT devices may of course be mobile devices. So, how do these assumptions affect privacy?

11.5.1 Brief Privacy Analysis

The IoT device will somehow need to verify that the accessing units have legitimate right to the IoT services. This usually implies entity authentication and some means of access control and authorization. Authentication, in general, is based on exchanged identities, and this immediately brings up identity privacy and all associated traits (including location privacy and untraceability). Access to an IoT service may not be all that privacy sensitive per se, but some services may be sensitive or sensitive when coupled with information such as the `<time,location>` of the access attempt. What we need to consider with respect to privacy is (as a minimum):

- Identity privacy of human user/proxy device
- Location privacy of human user/proxy device
- Traceability of human user/proxy device
- Transaction privacy (with respect to consumed services)
- Content (data) privacy

Traceability would imply that the same identity is used several times or that one may be able to link different identities to the same user. The above list is quite similar to the privacy issues we found for cellular subscriptions, but we note that cellular services are sufficiently generic that they probably don't by themselves reveal much in terms of "transaction privacy" (consumed services).

11.5.2 Security and Privacy for IoT

IoT devices all have an IP stack and it is reasonable to assume that the most common type of communications security protection is by means of standard

IP protection methods, like SSL/TLS [78], EAP [26] and IPsec [163]. So, adequate communication security is generally available.

What about privacy then? Well, most of the available methods are not really all that well suited for concealing the identity or address of the communicating partners. They are, on the other hand, well suited for data privacy and used appropriately may also protect against many transaction privacy threats. What the IP layer methods cannot protect against is information leaked on the lower layers. That is, identity information on layer 2, which would include MAC address for WLAN, will still be visible. The WLAN protection methods, like WiFi Protected Access (WPA) [139], is not well suited for this purpose either. That is, while WPA has its flaws [131], it will nevertheless provide adequate data privacy for the over-the-air link. However, it was not designed to conceal MAC addresses and so identity- and location privacy suffers.

11.6 Credible Privacy for IoT-Device Access

11.6.1 A Lightweight Privacy Enhanced Device Access Protocol

In the following, we argue the case for a lightweight device authentication protocol that authenticates a device/service class rather than an individual device. The devices in question are providing public services. The proposed protocol is an online protocol and it uses a pseudo-random temporary referential identity scheme to provide user privacy. The protocol is a variant of the protocol described in [175] and this section is directly influenced by that paper (by one of us).

11.6.1.1 Use-Cases

The following use-cases should serve to illustrate where our device/service-access protocol could be deployed.

- **Vending Machine**
  The vending machine is controlled by an IoT device. Customers may purchase many different items.
- **Automated Gas Station**
  Automated gas stations are commonly accepting credit card payment. One may also envisage a system in which payment is means of an `App` running on a mobile phone. The `App` has IoT access credentials and runs the device access protocol towards the IoT device controlling the gas station pump.

- **Wi-Fi Login Services**
  The user may gain access rights for a period (per day, per week, etc.) with the Wi-Fi operator. The access right could apply to all Wi-Fi access points within the operator network or it may be restricted to a region.

11.6.1.2 Defining the Requirements

The device access protocol must be flexible enough to cater for many different scenarios. It must also be secure, fast, reliable, and provide credible privacy.

- *Flexible service authorization during access*
  Flexibility is needed when defining what an "access right" amounts to. Time period or a defined number of access events? Specific device or any device? The service authorization may be handled by the IoT device or it may be handled by a remote server. We do not propose that authorization be part of the device access protocol per se, but rather that it be possible to offer flexible authorization in conjunction with the device access.
- *Online vs. Offline*
  Is the device operator to be online or offline during device access? If authorization is provided centrally (by the device operator) then clearly the device must be online during the authorization phase. We shall generally assume that the devices are online whenever needed (Assumption 1).
- *No Group Keys*
  We have stated that the access should be for a service type rather than for a specific device. However, we emphatically do not want a scheme with a symmetric group key for all devices with the given service type. Group keys are all too often a security and privacy liability. Initially, there may be some benefits to a simple en bloc key administration, but the cost and complexity of rectifying key compromise will be high indeed since it will a whole set of devices. We also strongly want to avoid a situation where the compromise of one device more-or-less automatically leads to compromise of other devices.
- *Credible Privacy* User privacy is a primary requirement and the user must be afforded credible privacy protection, in particular, for identity privacy, location privacy, data privacy (if applicable), untraceability and transaction privacy (with respect to consumed service). The privacy requirements are important toward external parties, but also internally. Specifically, we want to limit the IoT devices ability to compromise the privacy of the users. This requirement means that identities must be concealed and that the protocol payload must be confidentiality pro-

tected. Furthermore, should the service delivery be by means of data transfers (music/film/other contents) then these objects must also be protected (transaction privacy). The protocols must thus also provide key distribution/key agreement.

- *Preference for Simplicity and Performance*
  The device/service access protocol should be as simple and lightweight as possible with respect to computational and communication needs. A simple protocol is also simple to understand, and thereby simpler to verify. Perhaps even more important, a simple protocol is less likely to contain implementation errors.

11.6.2 Reference Architecture and Assumptions

11.6.3 Principal Entities

We will have three different types of principal entities and an intruder/adversary in our model:

- `OPR` : An operator entity which manages the IoT devices.
- `USR` : The user, represented by a proxy device.
- `DEV` : The generic IoT device, which will provide/facilitate services.
- `DYI` : The generic network-access intruder/adversary.

As is customary in the literature, we assume that we have honest principals. That is, the principal entities have good intentions and will faithfully conduct their business. In this sense, the principals can trust each other, but the trust must be enforceable (to provide trustworthiness) since we also have an intruder in the system (`DYI`). The intruder may, amongst others, masquerade as a principal entity. It is a primary goal for the authentication protocol to prevent the masquerade from being successful. More on our intruder in subsection 11.6.4.

11.6.3.1 Service Access Architecture

Figure 11.1 depicts the service access architecture. The user (`USR`) has gained access to a device (`DEV 3`), with the operator (`OPR`) being online (`B`-interface) during the authentication phase. We model a one-to-one correspondence between a service and the associated provider for simplicity. Multiple providers may offer the same basic service and one provider may offer multiple distinct services. One may also model different consumers, different services and different providers as distinguishable over the interfaces. However, for sake of simplicity, we only consider the minimal model as depicted in figure 11.1.

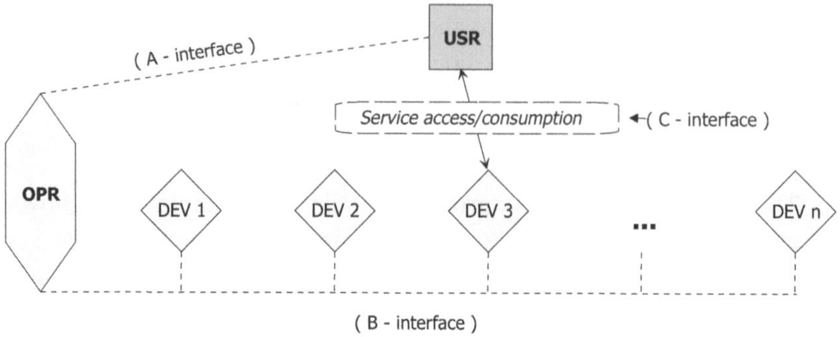

Figure 11.1 Simplified IoT Service Access Architecture

11.6.3.2 Interfaces and Channels

Between the principal entities we define the following interfaces:

- *A-interface: Between OPR and USR*
  The interface is used for all agreements between the user and the operator, including exchange of security credentials to facilitate service access between the user and the devices. This interface does not need to be operational during service access.
- *B-Interface: Between OPR and DEV*
  This includes device set-up, and it may include session authorization during user device access.
- *C-interface: Between USR and DEV*
  The C-channel is established over this interface during service access setup and will be available during service consumption.

Each interface is associated with a logical channel, which is named after the interface.

11.6.3.3 Security Assumptions

The A-channel is required to be an authenticated and fully secured channel. Security agreement for the A-channel is assumed to be pre-arranged. The concrete setup of the A-channel is not further defined here.

The B-channel is required to be an authenticated and fully secured channel. The operator will pre-configure the devices with device credentials during device deployment.

The C-channel will need to be mutually authenticated, with respect to service consumption rights. The C-channel must support data integrity protection and data confidentiality protection.

11.6.3.4 Computational Performance

We do not anticipate that careful and discriminate use of normal public-key primitives should be problematic. However, that assumption may be at odds with some resource constrained devices. Thus, there is a case for making the access protocol as lightweight as possible with respect to computational requirements. Still, we need credible and future proof security and so we cannot compromise by using weak security schemes. The natural choice then is to base the design around symmetric-key cryptographic primitives optimized for execution on resource constrained devices. This is not as limiting as it may sound; The AES [197] cipher primitive is optimized for performance and performance is also a key criterion for the new upcoming secure hash algorithm (SHA-3), expected to be selected by NIST by late 2012.

11.6.3.5 Trust Relationships

We define the following trust relationships:

- **Trust** OPR $\rightleftarrows$ DEV
  The operator has full security jurisdiction over the devices. Devices are plentiful and may become corrupted or may otherwise fail, and therefore the operator will only have qualified trust in a device. The devices have unqualified trust in the operator.
- **Trust** OPR$\rightarrow$USR
  This relationship is governed and limited by contractual agreement. The contract has a confined and definitive scope in time and in terms of what the operator is to provide to the user.
- **Trust** USR$\rightarrow$OPR
  The user must trust the operator with respect to respect service purchase and consumption. Privacy: The user will not need to fully trust the operator with information regarding service consumption (time/location etc), but must be prepared to trust the operator with payment information and possibly with identity information.
- **Trust** USR $\rightleftarrows$ DEV
  There is no *a priori* trust between these entities. All trust must be through operator mediation and the established trust depends on trust transitivity. Transitive trust is indirect and we must assume it to be weaker than direct trust. The trust level between the user and device is affected by whether the operator is online or not. For the offline case, neither party can be assured that the operator still authorizes the access.

Privacy: The user has no inherent reason to trust the device. The trust should be strictly limited to the needs for service access.

11.6.4 Intruder Model and Intruder Mitigation

The standard intruder model is the well-known Dolev-Yao (DY) intruder model [81]. The DY intruder can intercept, modify, delete and inject messages at will. The intruder has a complete archive of all previous message exchanges and it uses this to the full extent to attack the system. The DY intruder is also a privacy intruder and will exploit leakage of privacy sensitive data to the full extent. However, the DY intruder also has its limitations. For instance, the standard assumption is that it cannot physically compromise a principal and it cannot actually break cryptographic primitives. Real-world IoT intruders may not have the powers that a DY intruder has, but one should be careful about assuming the IoT intruder to be a weak intruder. The IoT intruder will also have powers that the DY intruder has not, including the ability to break "weak" crypto primitives and crypto system with too short keys etc. The IoT intruder may also, occasionally, be capable of physically intruding on devices and gain illicit access to the devices. It is vital that compromise of one or more devices will not give the intruder too much of an advantage in compromising other principal entities. Therefore, secrets and sensitive data (key material, identities etc.) stored at the devices should per se permit the DYI to compromise other principals, and this effectively rules out designs with shared symmetric-key key material. Cryptographically speaking, one should only use sound primitives and require the key sizes etc to be "sufficiently long". We advise adhering to recommendations such as the ECRYPT II recommendations [91].

11.6.5 Security Requirements

The user (USR) needs to verify that the device (DEV) is authorized to provide services (by the operator (OPR)). Correspondingly, the device (DEV) must be assured that the user (USR) has access rights to the indicated service. The operator (OPR) is assumed to be online during the initial device access. Should the B-channel be unavailable during device access, then the access protocol must provide assurance that the service access once was sanctioned by the operator. A suitable set of session keys must be agreed for the subsequent device access and service access procedures.

11.6.6 Privacy Requirements

The user identity, whether assigned, associated or emergent, should not be disclosed to any unauthorized parties. We emphasize here that unauthorized should be with respect to privacy; this means that for instance a local temporary identifier should not (automatically) be disclosed outside the local scope even for otherwise authorized parties. The user location should not unduly be disclosed and tracking of the user should not be permitted or possible. Data privacy and transaction privacy (`<consumed service, time, location>`) should be provided for most, if not all, services. User control and visibility over privacy options and actually provided privacy is important.

11.6.7 Offline or Online Model

To achieve satisfactory security for the case where the B-channel is unavailable is not easy to achieve with the requirements that the device need not know the identity of the user. To achieve the given goals, and to avoid using group keys, it is necessary to use asymmetric crypto primitives. There already exists several suitable public-key/digital certificate protocols and we suggest using one of these for the offline case. An example of an existing protocols is TLS (TLS v1.2, RFC 5246 [78]). Certificate revocation should be checked (RFC 3280 [136]), but in an offline model, this information may be obsolete.

If we model the operator to be online we can simplify the protocol structure and provide an assurance level that is not attainable with offline protocols [47]. The operator provides assurance and can forward key material to the device upon request, thereby removing the need for the user and the device to have complete credentials to start off with. This makes it feasible to carry out all operations with symmetric-key primitives.

11.6.8 PELDA – Outline of an Online Solution

The basic "Privacy Enhanced Lightweight Device Authentication" (PELDA) protocol [175] is depicted in Figure 11.2. The outline follows the description given in [175] in considerable detail. The PELDA protocol is only an example protocol and its purpose is mainly to highlight that it is possible to have simple lightweight protocols that nevertheless provide credible privacy.

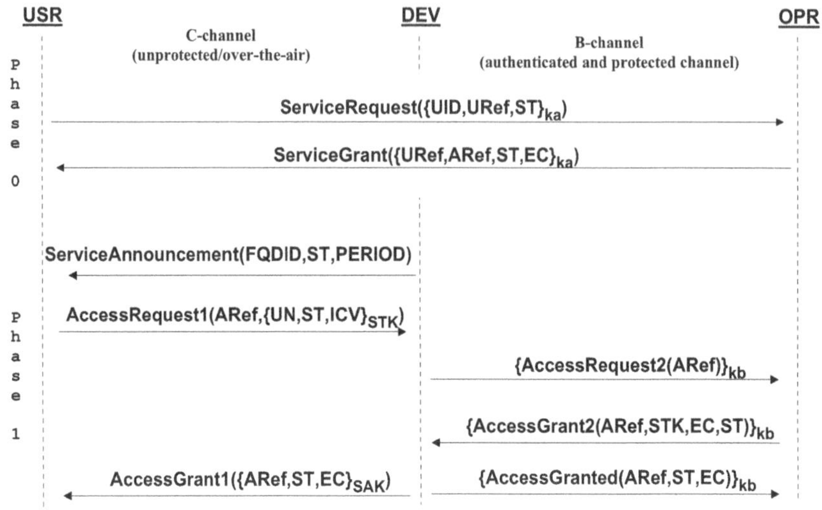

Figure 11.2 Outline of the basic PELDA protocol

11.6.8.1 PELDA Keys

In accordance with ECRYPT II [91], we advocate all keys to be at least 128-bit long. We additionally mandate that the identities, references and (pseudo-random) nonces be 128-bit long.

During execution of the PELDA protocol, several symmetric keys are derived. These are the "Service Type Key (STK)", the "Service Access Key (SAK)" and the "Temporary Access Key ($TAKx$)". The STK and SAK are key deriving keys. The TAK is used for protection of service content, and it consists of an integrity ($TAKi$) key and a confidentiality ($TAKc$) key. One could also have separate uplink and downlink key sets, but we have chosen not do so here.

11.6.8.2 The Service Type Key (STK)

The STK is derived for a specific user identified by the "user reference" ($URef$). The key derivation transform uses a symmetric-key encryption function, E, and the bitwise exclusive-or function. The $kdf1$ function has the property that even a party that possess the "input key" and the output key will not be able to deduce the input parameter. The "input key" in question is not an ordinary encryption key, but instead the service type identifier (ST). The input parameter $URef'$ is a MAC-modified pseudo-random user reference. The $URef$ is required to be confidential to all but the user (USR) and the operator (OPR). This use of encrypt-and-xor transform is not entirely new. It was also used in the MILENAGE authentication algorithm set used in cellular systems (UMTS and LTE) [18].

$$kdf1_{ST}(URef') \rightarrow STK \tag{11.1}$$

With the encrypt-and-xor part:

$$kdf1 : (E_{ST}(URef') \oplus URef') \rightarrow STK$$

11.6.8.3 The Service Access Key (SAK)

This key is specific to one device. The key derivation function takes STK as the input key. The input parameters, all non-secret, include the fully qualified device identity ($FQDID$), the service type (ST), a user generated pseudo-random nonce UN and the access reference ($ARef$). The $ARef$ is generated by the operator and must be guaranteed to be unique with respect to the request context (given in the `ServiceRequest` message).

$$kdf2_{STK}(FQDID, ARef, ST, UN) \rightarrow SAK \tag{11.2}$$

11.6.8.4 The Temporary Access Keys (TAK)

These are the session keys. The key derivation function takes the SAK as the input key. The only input parameter is the $PERIOD$ identifier (which is broadcast by the device). Whenever the $PERIOD$ indication in the `ServiceAnnouncement` changes the TAK keys must be re-computed. The expiry condition (EC) may not necessarily coincide with the $PERIOD$ announcement and expiry of EC will lead to expiry of the whole service context irrespective of the $PERIOD$ expiry.

$$kdf3_{SAK}(PERIOD) \rightarrow TAKi||TAKc \tag{11.3}$$

11.6.8.5 The User References ($URef, URef'$)

The $URef$ is a user reference generated by the user. It is assumed to be private to the user (USR) and is only shared with the operator (OPR). The $URef$ must be unpredictable and uniformly distributed. We assume it to be generated by a cryptographic pseudo-random number generator.

$$prf(\cdot) \rightarrow URef \tag{11.4}$$

The $URef'$ is an associated user reference and it is cryptographically bounded to one specific device. It is generated from $URef$ and the fully-qualified device identity ($FQDID$) with a MAC function, under control of the ka key. The ka key (for the A-channel) is only available to the USR and the OPR. The device in question does **not** have access to ka.

$$MAC_{ka}(URef, FQDID) \rightarrow URef' \tag{11.5}$$

11.6.9 PELDA Protocol Description

We now present the PELDA protocol in an augmented Alice–Bob notation. We assume the presence of a pseudo-random number function (prf), suitable (block cipher) symmetric-key primitives and a MAC function. We also assume that keys for the A-channel (ka) are agreed prior to PELDA execution and that keys for the B-channel (kb) are available during PELDA phase 1 execution. Figure 11.3 depicts the information elements (IE).

11.6.9.1 PELDA Phase 0 - Access Agreement

In phase-0, we have the user ordering the services from the operator. This phase may take place immediately before service consumption, but there may also be a prologed period between phase-0 and phase-2.

0. Preparations (pre-computation possible)
   - USR: $prf(\cdot) \rightarrow URef$
   - USR: Encrypt the ServiceRequest message with ka.

1. USR→OPR: ServiceRequest($\{UID, URef, ST,\}_{ka}$)
   - OPR: Decrypt the ServiceRequest message with ka.
   - OPR: $prf(\cdot) \rightarrow ARef$
   - OPR: Encrypt the ServiceGrant message with ka.

2. OPR→USR: ServiceGrant($\{URef, ARef, ST, EC\}_{ka}$)
   - USR: Decrypt the ServiceGrant message with ka.
   - USR: Extract the parameters.

By the end of phase 0, the user (USR) and operator (OPR) have agreed services (ST) and a validity period (EC). The operator knows about $URef$ and the user knows about the access reference ($ARef$), which the user must present to a device during access. The $ARef$ will be used in plaintext in phase 1 and it should, for privacy reasons, ideally only be used once.

11.6.9.2 PELDA Phase 1 - Access Request

Phase-1 is concerned with the access request and the agreement on suitable session keys. This is the main part of the PELDA protocol.

0. `DEV→all: ServiceAnnouncement(FQDID,ST,PERIOD)`
   - `USR:` $prf(\cdot) \rightarrow UN$
   - `USR:` $MAC_{ka}(URef, FQDID) \rightarrow URef'$
   - `USR:` $kdf1_{ST}(URef') \rightarrow STK$
   - `USR:` $kdf2_{STK}(FQDID, ARef, ST, UN) \rightarrow SAK$
   - `USR:` $MAC_{STK}(FQDID, PERIOD, ARef, ST, EC) \rightarrow ICV$

1. `USR→DEV: AccessRequest1(ARef,`$\{UN, ST, ICV\}_{STK}$`)`
   - `DEV: Store` $\{UN, ST, ICV\}_{STK}$
   - `DEV: Derive` $nonce$

2. `DEV→OPR: AccessRequest2(`$\{ARef, nonce\}_{kb}$`)`
   - `OPR:` $MAC_{ka}(URef, FQDID) \rightarrow URef'$
   - `OPR:` $kdf1_{ST}(URef') \rightarrow STK$

3. `OPR→DEV: AccessGrant2(`$\{ARef, STK, EC, ST\}_{kb}$`)`
   - `DEV: Decrypt` $\{UN, ST, ICV\}_{STK}$ `from AccessRequest1`
   - `DEV: Check:` $MAC_{STK}(FQDID, PERIOD, ARef, ST, EC) = ICV$
   - `DEV:` $kdf2_{STK}(FQDID, ARef, ST, UN) \rightarrow SAK$

4. `DEV→USR: AccessGrant1(`$\{ARef, ST, EC\}_{SAK}$`)`
   `DEV→OPR: AccessGranted(`$\{ARef, ST, EC, nonce\}_{kb}$`)`
   - `USR,DEV:` $kdf3_{SAK}(PERIOD) \rightarrow TAKi||TAKc$

We note that the $\{UN, ST, ICV\}_{STK}$ block, in step 1, is not understood by the device. Instead the device just forwards $ARef$ to the operator. For cryptographic reasons, we add a $nonce$ to the message; its sole purpose is to make crypto-analysis harder. The $ARef$ is used by the operator to retrieve the access context and compute the $URef'$ and the STK. The operator then forwards the STK key and this allows the device to decrypt the block received in step 1. The device can then verify the parameters, generate session keys and confirm the service to the user. The device also notifies the operator that the access was granted. The $nonce$ in the final message is there to avoid certain possible pitfalls with two identical messages being encrypted with different keys. The $nonce$ should be the same $nonce$ as was used in the `AccessRequest2` message.

```
----------------------------------------------------------------------
UID   : Globally Unique Permanent privacy-sensitive user identity;
DID   : Globally Unique Permanent public device identity;
SID   : Permanent public server identity;
FQDID : SID||DID; Fully Qualified Device Identity;
URef  : User Reference; Also URef';
ARef  : Access Reference;
UN    : User Nonce;
ST    : "Service Type" identifier;
EC    : "Expiry Condition" identifier;
ICV   : Integrity Check Value;
STK   : Service Type Key;
SAK   : Service Access Key;
TAKx  : Temporary Access Key x; x={i|c}, (integrity|confidentiality)
ka    : Pre-shared secret, known to the user and the operator;
kb    : Pre-shared secret, known to the device and the operator;
nonce : Generic nonce;
----------------------------------------------------------------------
```

Figure 11.3 PELDA - Overview over the Information Elements

11.6.9.3 PELDA Phase 2 - Rekeying

Re-keying of $TAKx$ takes place when the $PERIOD$ identifier in the broadcast message `ServiceAnnouncement` changes. The transferred payloads over the `C`-channel should include a $PERIOD$ indicator to allow seamless key changes and uninterrupted services.

11.6.10 Abridged Protocol Analysis

11.6.10.1 Complexity and Efficiency Aspects

The computational cost of the PELDA protocol is modest and only symmetric-key primitives are used. Even devices with limited computational power will be able to run the protocol without any problem. With a cipher block size of 128 bits all messages will comfortably fit within 4–5 blocks. This makes the communications overhead modest and we expect all IP-enabled devices to tackle this quite easily. With respect to round-trip delays for phase 1, a full round-trip from the user, via the device and to the operator is required. Observe that `AccessGrant1` and `AccessGranted` may be sent in parallel. Overall, the conclusion must be that the PELDA protocol truly is a lightweight protocol.

11.6.10.2 Informal Security Analysis

For PELDA phase 0, we have the pre-condition that the A-channel is authenticated and protected. We may therefore safely assume that security is maintained for the A-channel.

For PELDA phase 1, we have that the user initially generates the $URef$ and UN elements, which are pseudo-random and which are unique and unpredictable. The STK is derived by the user and is known to the user to be a fresh secret key. The $ARef$ is known to the user to be associated with the fresh $URef$ (phase 0). Given the trust relationships and security control (the operator has *security jurisdiction* over the user) stated earlier the user can afford to believe that the $ARef$ is fresh and unique.

In `AccessRequest1`, the user sends $ARef$ to the device. The device forwards $ARef$ to the server over the authenticated and fully protected the B-channel. The device fully trusts the operator (the operator has *security jurisdiction* over the device too). Therefore, when the device receives the `AccessGrant2` message it has assurance that the $ARef$ is valid and that it is associated with STK, EC and ST. By verifying the ICV the device also has assurance that the access attempt is indeed for it, for the $PERIOD$ (timeliness) and for the user (indicated here by $ARef$).

Subsequent to `AccessGrant1` the user has assurance that the device was recognized by the server through the use of SAK. (SAK can only be derived by a party which knows STK and $ARef$). The user already has assurance of STK and $ARef$, and it therefore accepts the device as being valid.

The operator does not get explicit assurance of the user during PELDA phase 1, but relies on the device to ascertain the user ($Aref$). Nevertheless, the operator has assurance that the device (in `AccessGrant2`) will only provide services as agreed for $ARef$ (encoded in ST, EC), and so it has covered its needs.

11.6.10.3 Informal Privacy Analysis

Our main privacy requirement is that neither the device nor any external entity should be able to learn the user identity. Since we have that neither the device nor any external entity ever actually sees the user identity in plaintext (neither the UID nor the $URef$), we may conclude that the requirement is fulfilled.

The $ARef$ is required to be known to the device and it is also transmitted in plaintext. Frequent re-use of the $ARef$ could lead to it becoming an emergent identity for the user. The device is a semi-trusted entity and we don't worry too much about the fact that device is able to track the user for the lifetime of $ARef$. However, external entities should clearly not be able to track the user. Therefore, one must assure that the $ARef$ is not exposed (`AccessRequest1`) too many times. For this reason, the user client (at the proxy device) should keep track of the $ARef$ exposure.

If we assume prudent use of $ARef$ there should not be any information pertaining to the user identify being divulged during PELDA execution. Still, one must be aware that certain service types may lead to distinguishable payload patterns and this may allow an intruder to deduce the service type and the time/location when the service was consumed. Should there be additional distinct clues, then the intruder may recreate more context and one cannot entirely preclude this by means of the access security protocol. One may, of course, introduce dummy payload traffic etc. to mask the payload patterns, but this is not as easy as it appears and it cannot be done without introducing communications overhead or delays in service provisioning.

11.6.11 Concluding Remarks on the PELDA Protocol

We have presented the privacy enhanced lightweight device authentication (PELDA) protocol. The goal of the protocol was to facilitate access to publicly operated IoT-based services such that the user may focus on service access rather than on device access. Privacy is the main design motivation for the PELDA protocol. One concern has been that inexpensive and widely distributed IoT devices cannot be expected to provide the best protection for privacy sensitive data. The PELDA protocol was designed with this in mind and it is optimized to avoid storing privacy sensitive data at the devices. The protocol uses a disposable "access references" in place of a permanent identity, and it is therefore not that much sensitive information that the device can leak/divulge. User privacy will suffer if $ARef$ is re-used for a prolonged period. A potential drawback to the PELDA scheme is that it requires the operator to be online, but IoT devices are naturally expected to be connected most of the time. (Ref. Assumption 1). The bare-bones PELDA protocol is a simple protocol with few round-trips, a relatively small payload and modest crypto-performance requirements.

The other IoT assumptions are also addressed by the PELDA protocol. It may appear that Assumption 5 (user knowledge of the device identity) is not taken into account, but the fact that the device broadcasts its (claimed) identity over an IP-enabled channel doesn't really mean that the human users are aware of the identity.

11.7 Summary

In this chapter, we have investigated privacy in the context of IoT services and devices. For our investigation, we made a set of assumption on what IoT is and how the user relates to IoT devices.

One major concern that we addressed in this chapter was that of trust and trustworthiness. The treatment was given in the context of IoT, but the trust/trustworthiness aspects are of course generic in nature. In particular, we tried to convey the notion that security tends not to be the end-goal but rather a means to support trust and, in particular, to ensure that the trust can be enforced (trustworthiness). Privacy and trust are both higher layer concepts and are to some extent intermingled in the sense that credible privacy instils increased trust and that trust/trustworthiness may be regarded as a prerequisite for beliefs in attained privacy.

The chapter concluded with presentation of the proof-of-concept "privacy enhanced lightweight device authentication" (PELDA) protocol. The protocol is designed to be efficient and to provide credible and practical user privacy. Identity privacy is provided through use of user-defined (possibly one-time) pseudo-random identifiers. Location privacy is also provided to the extent that there is not discernible connection or association between the permanent user identity and the identity used at the device access. Thus, an intruder/adversary may learn that there is somebody accessing services within an area, but given that the intruder can only associate a (one-time) pseudo-random identifier with the access the intruder essential learns no privacy sensitive information at all.

The PELDA protocol is only a proof-of-concept protocol, but it does demonstrate that one can indeed have fast, efficient access protocols that take privacy into account. The protocol shares some traits with the protocols proposed in Chapter 8, and these protocols share the use of pseudo-random temporal identities to provide location- and identity privacy. So privacy-by-design (Section 1.5) is indeed possible. More work is still needed in this area and we note that a secure, fast, efficient and clean design does not guarantee adoption in the market place.

12

Privacy in the Cloud

Telecommunication industry as a key provider of connecting networks can play a key role in the Cloud marketplace. It is already recognized that cloud-enabled businesses will generate substantial profits in a near future [111]. In this chapter, we outline the long-term view of privacy and investigate the essential location independence inherent in cloud computing, and examine the associated service and user mobility aspects. We discuss the idea trust-based approach to privacy protection. As a case study, we address privacy implications associated with problem known as the term-of-service threat.

12.1 Cloud Characteristics

In this section, we briefly outline some key characteristics of the cloud with respect to trust, security and privacy.

12.1.1 Location Independence and Mobility as a Basic Premise

A central characteristic of cloud computing is that the services are essentially location independent. Furthermore, we claim that "service mobility" during service production/consumption is also an aspect that must be viewed as a premise. That is, the cloud service (i.e. the VM) might be migrated physically during service production[1].

The cloud service consumer (user) is also location independent, and if we assume a wireless/cellular user, then it is clear that the user may move during service consumption. Thus, we claim that there can be no *a priori* assumption about the location of neither the cloud service producer nor of the cloud service consumer. Furthermore, the location of both may change during the production/consumption of the service(s). The above insights are impor-

[1] This should be entirely transparent to the user and it should likely also be transparent to the service itself.

tant since the mobility may affect the privacy of the cloud service consumers adversely.

12.1.2 Lifetime of Security and Privacy Protection

When designing cryptographic primitives, security protocols and privacy preserving methods one must make assumptions about the perceived lifetime of the objects to be protected.

12.1.3 Communications Security

Cloud computing relies heavily on communications capabilities and so one needs to consider "communications security" protection. The lifetime of "communications security" protection is often in the order of the duration of the session in question. So we are dealing with lifetimes ranging from $\leq$ 1 second and up to possibly a few hours. To allow for safety margin, we may require the protection to be effective for 1 year. We do not assert that the protection needs are limited to these lifetimes, but we claim it to be outside the scope of communications security to provide a higher level of protection.

12.1.4 Device Security

The lifetime of "device security" protection is normally in the order of the lifetime of the device itself. For personal gadgets (laptop, mobile phone, mp3 player etc), this is in the order of 2–5 years. The cloud analogue of a "device" is a single unit of the service execution environment, i.e. a server unit. We estimate the lifetime to be up to 10 years.

12.1.5 Data Storage Protection

The lifetime of "data storage protection" can be longer than the lifetime of the storage media[2]. It can also outlast the protection methods and the key material. What we have then is that protected objects will routinely need to be migrated onto new storage media during its lifetime. Key material will expire, and even the protection methods and encryption primitives are likely to be outdated during the protection period. Thus, it must be possible to refresh the key material, negotiate new algorithms and cryptographic primitives for the protected objects.

---

[2] The storage media can be considered a "device".

12.1.6 Archival Storage

Archival storage is where the lifespan is assumed to be very long ($\geq$ 30 years). We assume that archival objects will need to migrate between storage media and that protection methods will change during its lifespan.

12.1.7 Personal Privacy

The lifetime of a privacy object could potentially be that of the data subject itself. That is, we can potentially have protection lifetimes of 100 years or more.

12.2 Security and Privacy Aspects

12.2.1 Security Aspects

In this section, we overview essential aspects of security and privacy that should be considered in cloud setting.

12.2.1.1 Service Access Security

This is a "communications security" case. By and large the security community has adequate solutions for this area, even for wireless/cellular access security where identity and location privacy is problematic [169, 179]. There is still room for improvements, particularly in the area of identity structures and identity management.

12.2.1.2 Service Execution Security

This is partially a "device security"' case and partially a "data storage protection"' case. There are many challenges in this area, but also solutions. The professional high-assurance cloud service provider can afford the best security surveillance, the best firewalls, the best anti-malware protection and the best intrusion detection available in a 24/7 operation. The use of visualization technologies seems to be beneficial to security in an environment with professional operations and with proper security precautions in place [30, 66]. Thus, many of the traditional computer security problems can be effectively contained and mitigated.

12.2.2 The Right to Privacy

The right to privacy is fundamental [248], but yet it is not a right set in stone. Popular opinion varies over time and the right to privacy is subject to societal acceptance and balanced against other needs[3]. This plasticity affects privacy since one cannot expect the laws or the business practices to remain invariant for the duration of the privacy protection period.

12.2.3 Privacy Aspects

As we have discussed in the previous chapters, privacy is different from security in several aspects. One aspect is the strong asymmetry in the interests of the principal parties. For instance, if a user has sensitive medical information about himself/herself, then he/she may not want to disclose that information. However, to the cloud service provider, the medical information may be a valuable piece of information since it may be sold to, for instance, insurance companies or to pharmaceutical companies etc.

Another aspect of privacy is the "*once exposed - forever lost*"' property. In security we also have problems with exposed secrets, but more often than not it is possible to recover from the situation. It may incur economical losses, loss of secrecy and loss of reputation, but overall the losses are temporary and fully recoverable. Not so with privacy, which rather has an "entropy" type of property in the information theoretic sense. That is, information leakage can only increase over time and while one can limit the leakage one cannot truly prevent it.

12.2.4 The Long-Term View of Privacy

Here we assume that there is information leakage and that this inevitably will affect privacy sensitive data. Given that we deal with "personal" privacy it then follows that solutions that attempt to protect the privacy must be made with a life-time perspective. To this end, any viable solution must take the view that privacy leakage must be contained for the lifespan of an average person.

The additional problem one faces is that one may not *a priori* know which data are sensitive since non-sensitive data may become sensitive in the presence of other (potentially also non-sensitive) data. We, therefore, assume that data correlation (for example, by data mining) will invariably lead to a higher degree of privacy sensitivity.

We also know that information leakage is inevitable in some respects. As demonstrated by Ristenpart *et al* [217], cloud computing and VMs may be

[3] As an example, perception of the right to privacy changed noticeably in the US after the 9/11 attacks.

location independent in theory, but it is often possible to deduce the mappings to physical locations. Of course, if the access is by wireless/cellular methods then one may also routinely expose the identity and location of the user [179].

12.2.5 Privacy and Data Exposure

Much, if not most, private information is not privacy sensitive per se. Most people routinely and voluntarily expose private information to external parties. This is often done for convenience (online shopping registration etc) and often due to personal preferences (facebook entries etc). It may be questionable whether most people understand the longtime implications of this, but it is a fact they nevertheless voluntarily expose potentially privacy sensitive data [180].

This, however, does not mean that other parties that have retrieved personal data have the right to expose the very same data. Here data ownership comes into play, and one must legally have the "right to expose" data in order to do so. There are grey areas, like the right to use anonymized data or to use aggregated (and sufficiently anonymized) data[4]. Therefore, we assume that a cloud service provider does not have the "right to expose" data unless this right has been explicitly obtained.

12.3 Principal Parties in the Cloud

12.3.1 Individual Users

There will be a huge number of individual users and each user will have very limited negotiation power with respect to the cloud service provider. There is, therefore, a fairly strong asymmetry in this respect.

12.3.2 Aggregate Users

By an aggregate user we mean a company, a corporation or an organization that manages a set of users. The aggregate users have a common management authority and it is this authority that negotiates service agreements etc with cloud service providers. The asymmetry in size and negotiation power is now greatly reduced.

[4] The anonymization must, of course, be effective for the privacy lifetime.

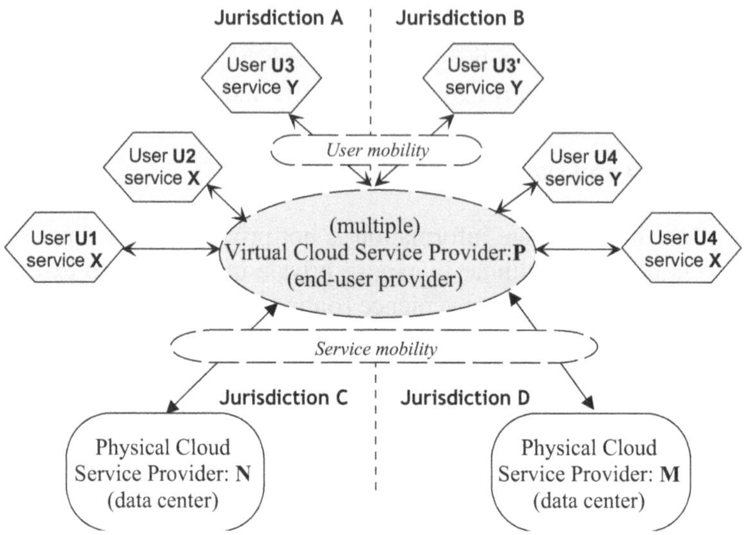

Figure 12.1 Architectural Outline

12.3.3 Cloud Service Provider

12.3.3.1 Taxonomy

We may classify a cloud service provider to be either *public* or *private*. We may further divide the service providers into *virtual service providers* and *physical service providers*. This distinction is inspired by the distinction one makes for mobile service providers A generic cloud service provider would provide both virtual and physical services.

12.3.4 Reference Architecture

Figure 12.1 depicts the generic cloud architecture model for our case. There are several jurisdictions (state/country) boundaries, there are many users and services. We also have multiple physical service providers and multiple virtual providers.

12.3.5 Jurisdictions and Legal Issues

We note that both the location independence and mobility may cause a service or a user to cross jurisdiction boundaries. This can have privacy implications since the legal framework for privacy differs substantially around the world. As noted in [30], this may cause problems and could lead to weakened privacy.

12.3.6 Private vs. Public Cloud Service Provider

The *public* provider offers its service in the market. The *private* provider would not offer its services externally. It would typically be an in-house provider at some large company or organization.

12.3.7 Cloud Service Providers

We define a virtual cloud service provider (vCSP) to be a service provider that accepts cloud service contracts from user, but which does not physically have facilities to run the services.

Ultimately, the cloud services must be run on physical equipment. A physical cloud service provider (pCSP) may host both private and public services on the same physical infrastructure. It is conceivable, as a special case, that some services requires exclusive hardware access. This special case would then be reminiscent of providing services at a traditional data center.

12.3.8 Cloud Intruder

The classical intruder model is that of the Dolev-Yao Intruder (DYI) [81]. The classical DYI model is for communications security, but it can easily be applied to privacy as well. In this model, we have an intruder that will learn about all disclosed data and will be able to use it to the greatest possible effect. A cloud DYI is by definition capable of data mining, thus, the cloud DYI will automatically correlate privacy sensitive data to the greatest possible effect.

12.3.8.1 Dishonest Principals

The classic DYI cannot corrupt honest principals, but are the principals necessarily trustworthy and honest? We note that an initially honest principal party must remain honest and trustworthy for the duration of the protection period. This is no small requirement and it requires a thorough and well thought out data management strategy as well as suitable (and stable) legislative measures.

12.3.8.2 Abandoned Devices

Part of the "trustworthy" attribute of a principal is long-term data management, and this extends to include handling of discarded devices. We assume that the DYI can learn data on discarded and abandoned devices unless the media is fully destroyed or the data is cryptographically protected (archival storage protection).

12.4 Service Mobility and De-Correlation

12.4.1 Service Provider Mapping

The mapping of vCSP services on pCSP equipment allows for a level of indirection over the "service mobility" layer (Figure 12.1). What we have is that the vCSP must know (and authenticate) the user identity (*uid*) and it will have a reference to the virtual service sessions (*vVMref*). The vCSP should, under normal operation, have no need for knowing the location of the user. There should be pre-established trust relationships between the user and the vCSP.

The pCSP will need to know the identity of the vCSP and there must be a common reference (*pVMref*) to VMs executing on the pCSP platform. However, as long as the pCSP has assurance of the vCSP it has no (functional) need for knowing the user identity (*uid*) or the reference to the virtual session (*vVMref*)[5]. Obviously, the vCSP and pCSP must have a level of mutual trust in each other.

12.4.2 De-Correlation

We assert that the service mobility indirection allows for a measure of anonymization and de-correlation, in particular, with respect to identity- and location privacy. By proper use of pseudonymous identities and references this type of de-correlation is possible, as exemplified by the pseudo-random context reference identity scheme proposed in [169]. The Achilles heel of this type of scheme is that it requires the vCSP and pCSP to fully respect the privacy rules.

De-correlation at the service mobility layer can also be achieved by MIX functionality [57]. However, MIX technologies and Onion Routing of VMs may not be efficient solutions and they would require another party to be present (which introduces yet more trust dependencies).

[5] Likewise, the user would not normally have a need for knowing the pCSF location/identity or the physical VM reference (*pVMref*). However, the user may need assurance that the service is executed within a specific jurisdiction.

12.5 Trust in the Cloud

As noted by Chen and Sion [60], it is economically infeasible to compensate for lacking trust in a cloud service provider with encryption techniques alone. Thus, in this section we investigate using trust and trust relationships between the principal parties to enhance privacy.

12.5.1 Trust-based Privacy Protection

In this section, we consider an alternative approach to reduce privacy treats and enhance users confidence with respect to cloud applications [202]. We assume that completely trusted providers are not considered by users as a privacy treat. However, in the public cloud most providers will be trusted only to some degree [127]. The level of trust will vary during the lifetime and depend on context defined by provider location, known history of interactions, jurisdiction, time in business, size, etc. It will also depend on what kind of services are provided (for example, data storage, data mining, scientific computations, etc). All these parameters will create the context in which the user and service provider interact.

Reputation on keeping private data protected against leakage and reputation of behaving according to agreement will be based on history of interaction with service provider both by this specific client and also other clients whose opinions are considered as trustworthy.

The use of trust measurement to enhance privacy protection is based on the assumption that clients will have better privacy protection if privacy sensitive data are exposed to a more trustworthy service provider comparing with what may be expected from less trustworthy service providers.

As was explained in the previous sections and shown in research, long-term privacy cannot be protected by only cryptographic methods [244]. It is not feasible from cryptographic or economic points of view. Therefore, we propose to enhance privacy protection by monitoring trustworthiness of service providers. However, such approach should be combined with cryptographic and legislative means whenever it is possible.

Anonymization is an approach that can be used to achieve the needed levels of privacy protection when data are exposed to service provider with limited or unknown trustworthiness. Different approaches to achieve anonymity have been described in the literature [116]. In most cases, it is pseudo-anonymity and the degree of anonymity depends on trustworthiness of involved third parties.

Assume that there is a set of anonymizers $A = \{a_{i,j}|i = 1, \ldots\}$ that provide anonymization of various data (identity, location data, traffic data etc.). Some of the anonymizers can be run by users themselves and be trustworthy, but they can also be cloud-based anonymizers. A user s will interact with some anonymizers from A before exposing data to the cloud (that is, removing personally identifiable information from data). However, privacy protection will partly depend on how trustworthy anonymizing services are. Therefore, a user (subject) should be able to measure and continually monitor trustworthiness of members of A.

Let $\omega_{a_j}^{s_i}$ be an opinion of a user s_i about trustworthiness of a_j. The opinions, expressing trustworthiness, can be naturally expressed in terms of subjective logic [154, 155] as it was discussed in [202] where trustworthiness is expressed by opinions. These opinions may vary over the lifetime and depend on opinions of other users from S, where $S = \{s_i|i = 1, \ldots\}$, that have used anonymizers before. Each user s may change his opinion on a specific anonymizer from A. When it takes place, s broadcasts its new opinion to all members of his community S. Then each member of S updates relevant opinions in his database (more details can be found in [202]). The level of anonymization will depend on trustworthiness of an anonymizer. It is possible that some anonymizers may collude to disclose users identity. The concept of k-anonymity [67] in this context means that de-anonymization will require cooperation of at least k anonymizers.

We see the cloud as a set of cloud service providers $C = \{c_1, \ldots, c_k\}$. We assume that there are opinions on trustworthiness of services from C, denoted $\omega_{c_j}^{s_i}$, are also available (where $\omega_{c_j}^{s_i}$ is an opinion of a user s_i about trustworthiness of c_j).

One approach to increase privacy protection would be to split data into separate shares/parts and ask for anonymization by different anonymizers from A. Such splitting into shares itself can be seen as (partly) anonymization and can be done by users within their private clouds. A user may ask data anonymization service(s) to anonymize data and then submit it via a location anonymizing service to the cloud. Based on opinions about trustworthiness of the anonymizer and the cloud service provider, each user will decide on what level of anonymization is the most suitable.

We assume that different kinds of data have potentially different privacy threats. For example, a family name is more privacy sensitive than a first name. Protecting the most unique identifiers (with lowest k in the sense of k-anonymity) can help to increase privacy protection. Therefore, data that will be exposed to cloud service providers will have different levels of sensitivity.

It is difficult for users to measure such sensitivity since data that does not appear sensitive today may be very sensitive in the future. In the next subsection, we describe how trustworthiness of cloud service providers including anonymizers can be measured. For more details related to subjective logic, the reader is recommended to consult [154, 155].

12.6 Privacy Enhanced Cloud Services

Assume that r denotes data owned by a user. The user wants a cloud service provider to perform some service that involves exposure of r. If r contains privacy sensitive information, it may be used to violate user's privacy. Let $\pi(r)$ denote privacy sensitivity of r, where $\pi : D \rightarrow [0, 1]$ and D is a domain of user data that may be handled in the cloud. Privacy sensitivity function π is defined by a user and depends on state of the art and user's experience. We assume that π is given.

When a user considers cloud service providers $c_1, \ldots, c_k$ to handle r, he/she considers the trustworthiness of each of them. Less trustworthy service providers can be considered as being greater threat to privacy. Therefore, a user either selects a service provider that is trustworthy enough to handle data r with current privacy sensitivity $\pi(r)$ or will try to reduce privacy sensitivity for r. Privacy sensitivity of r can be reduced by different means such as encryption, anonymization, splitting in separate shares, etc.

Assume that the user plans to use c_j. Let $\alpha_i(r)$ denote anonymization by anonymizer a_i. If a_i is trustworthy, then $\pi(r) > \pi(\alpha_i(r))$, that is, anonymization reduces data privacy sensitivity. However, revealing data r to anonymizer a_i assumes that a_i is more trustworthy than c_j. After that, data can be sent $\alpha_i(r)$ to the cloud service provider c_j.

12.6.1 Privacy implications of Terms-of-Service

In this section, we consider as a case study, potential solutions to the problem now known as the *terms-of-service* (ToS) threat [226]. This threat emerges when the terms of service is formulated such that it may be changed unilaterally by the service producer. This can cause conflicts as was highlighted by the resent Instagram debacle [238], where Instagram changed its terms of service to allow itself the right to sell user photos. In this case, Instagram was forced to revise the change to its ToS after a public outcry, but it turns out that many of the "free" internet services do contain similar terms of service.

We specifically investigate the ToS threats that may arise in the context of free file storage and synchronization services. We analyze the power relationships between the service producer and the service consumer and we investigate the conflict of interests inherent in these agreements using the Conflicting Incentives Risk Analysis (CIRA) model [215].

We propose to address the ToS threat by a combination of several measures in what we call the *Umbrella Architecture*. The main part consists of a client-side *presentation layer security manager* (PLaSM) which will protect the user content. The PLaSM will provide a set of security protection features and these will be available at all times for the user. These will, as a minimum, include file encryption and file integrity services.

The other components are a *policy monitoring agent* (PoMA) and a *reputation manager* (ReMa). The PoMA will monitor the ToS and will alert the consumer in case of changes and the reputation manager will gather information about the reliability and trustworthiness of the SP. The PoMA and the ReMa may also trigger actions, called strategies in CIRA parlance. These actions will then be carried out by the PLaSM. The net result will be improved security and privacy for user content stored by cloud file storage services.

12.6.2 Analysis of the Terms-of-Service Threat

Needless to say for a free service, but the offering party obviously gets to dictate the premises of the ToS. There is, of course, a larger societal context that the SP must take into account, both with respect to legal requirements and with respect to customer reactions. In [48], the authors discuss the legal aspects of ToS in the context of an ISP. The paper is US centric, but the general conclusions should hold in most jurisdictions.

The agreements are written in a language intended for lawyers rather than laymen. Furthermore, the SC generally has little specific knowledge about his/her rights to start with. So, informally, we summarize the ToS threat as the following:

- ToS statements are not very readable to the layman (read: ToS are seldom read).
- ToS are subject to unilateral change by the SP (and the notice could well go unnoticed).
- ToS regulate rights to the contents stored by the SP, and this may seriously affect the SC content ownership rights and SC privacy.

12.6.3 Accountability and Availability

Accountability is an important aspect for almost all cloud-based services. Big service producers like Apple, Google, Microsoft, Amazon, Dropbox etc. are generally reputable organizations, and we may assume that they will provide reasonable services and be accountable for them. Smaller outfits may or may not be reputable, and in the wake of bankruptcy or similar it will be anybody's guess as to how well behaved they will be. However, the catch is *playing by the rules*, which are those captured in the ToS. As demonstrated by the Instagram case, the companies exist in a societal context, and this may prevent a company from abusing their rights.

Availability is another important aspect. It is, of course, purported to be a major benefit of cloud storage and for the file synchronization services one will usually have local access to the data anyhow. In our context, we have decided not to pursue availability further, but note that this aspect has been addressed in the context of paid-for services [46].

In this context, we do not address the accountability and availability aspects directly. That is, we assume that the required minimum of accountability and availability features are already present in the provided services.

12.6.4 Privacy, Identification, Authentication, Integrity and Confidentiality

The identification used in most of the services is based on email addresses. These are not particularly privacy sensitive, although we want to avoid unnecessary exposure. Identity privacy itself is a concern, although not the primary privacy concern here. We note that the authentication schemes for free service seems not to be particularly strong, but with additional data protection they may be adequate. This needs to be verified. Otherwise, we note that privacy has many facets and that one correspondingly must have a multifaceted approach when counteracting and mitigating privacy problems [202].

In [229], the authors outline a set of security issues in delivery models in cloud computing. These issues are mostly concerned with enterprises using paid-for cloud services, but we note that many of the concerns are similar. Another paper which investigates these issues is [256] and this paper is interesting in that it distinguishes between different deployment models. With respect to our case we note that the consumer will not have much influence upon the chosen model, apart from what can be implemented at the client side. Another interesting paper is [38] in which there is an attempt at defining service level agreements (SLAs) for cloud security. Our context is different and so the consumer will not be able to negotiate SLA arrangements, but is rather left with a ToS that he/she cannot negotiate.

It should be evident that we need both data integrity and data confidentiality. The SC needs to ensure that the stored data isn't manipulated against his/her will and likewise the SC needs assurance that the stored data isn't unduly exposed. Many, if not most, of the free public file storage services do not offer data confidentiality services. Data integrity is offered to the extent that this is directly supported by the file systems used. The quality may be acceptable for most uses, but fails to cover cases where the SP is the source of the threat.

We also want access control for our data. The default should be that the SC is the only one with access. Other parties may be granted full or partial access by the SC. The service producers commonly provide schemes to allow this kind of access control. The authentication provided seems generally to rely on passwords and it seems only to be unilateral. We have not assessed the strength of the schemes, but suffice to say that they are designed for "low grade" systems. That is, they are probably fairly weak, but may still be statistically adequate for the given purpose. We, therefore, propose to rely on the existing authentication and access control scheme offered by the service producer, but we do not exclude the possibility of enhancing it. Requirements (partially fulfilled):

- Identity privacy (not addressed)
- Data confidentiality (strong requirement)
- Data integrity (partially fulfilled)
- Consumer–Producer Authentication scheme
- Access control (acceptable)

To summarize, we *must* provide data confidentiality and we ought to provide data integrity services. When it comes to the scope of the protection it should be evident that file data must be protected, but there is also a strong case for protecting file system data, particularly the file/directory names. One may provide enhanced access control, but there will be a cost to doing this. One may also improve the authentication, but it may be the case that the access oriented authentication is sufficient when one consider a scheme in which the data is explicitly protected independently of the access procedures. The two last requirements above will, therefore, largely be for the existence of an adequate solution.

12.6.5 Trust Aspects of the Consumer-Producer Relationship

The legal/contractual relationship is defined by the ToS, but what about the trust aspects? There are ways of assessing trust in an online context. We approach such as the one laid out in [210] with automated evaluation of Q&A sessions over the so-called online social networks (OSN). Reputation is keyword here and there are formalized approaches that directly use reputation in the model [156]. A survey of relevant proposals for handling trust and reputation in an online context is found in [157]. Of course, there are many aspects to trust. In [176], the author discusses this in the context of publicly available IoT services, and while the cloud context is somewhat different from an IoT context, many of the same ideas of trust in an IoT environment will also apply to cloud services. This leads us to assume that trust in free public cloud based services will largely be based on reputation and association with well-known brands. This is not the soundest basis one can have for security, and so we must provide some means of enforcement for the trust be warranted. Another survey paper handling trust and trust management in an internet application context is found in [126]. The context is not specific to cloud services, but the discussion is relevant in that it handles trust and decision making for internet applications.

12.6.6 Control Aspects of the Consumer-Producer Relationship

Who has control in our Consumer–Producer relationship? Given the asymmetry in power regarding the ToS it should be clear that SP obviously has both jurisdictional control and operational control over the service. That is, the control is there as long as the consumer continues to use the service. There are several free file storage services in the market and there is, therefore, a certain amount of competition. The consumer may have concrete needs, but he/she can nevertheless choose between different alternatives. However, once we have actually made a choice there will be transaction costs to switch to an alternative service. We should add to this picture that many services comes in bundles and as pre-installed software on smart mobiles, tablets and laptops. So we may conclude that the consumer has a great deal of power to choose, but that the ability is impeded by imperfect knowledge, by default (pre-installed) solutions, by technical inability to change configuration and by loss of convenience and cost optimizations.

12.6.7 The Conflicting Incentives Approach

The consumer and the producer have different perspectives, and this can also be expressed in terms of what kind of threats they pose to each other. One way to analyze Consumer–Producer relationship is to investigate the underlying incentives.

12.6.7.1 Conflicting Incentives Risk Analysis

The Conflicting Incentives Risk Analysis (CIRA) [215] is a method for analyzing risk under circumstances where it is hard to assess incident probability. This may be the case for infrequent events and generally for circumstances where past history cannot be used to predict future likelihoods. The CIRA method will instead assess the motives and incentives of the different principals. To do so the method draws on game theory, economics, psychology and decision theory. When analyzing the ToS threat scenarios we may benefit from using a similar approach. Technical threats by external intruders are, of course, still a major concern, but what if the motives and incentives of the principals are themselves a main driver behind many of the threats?

12.6.7.2 The Utility Function and the Strategy Concept

In CIRA, one has defined a risk owner and it defines the perspective taken. To our end we define the SC as the risk owner. The SP is the other principal entity. Then we have the strategy concept. A strategy here is some action that is intended to influence the utility function. The strategy owner is the principal that is in a position to execute the strategy. In our case, we want to define strategies that lower the threat against stored content and the utility function must correspond to these goals. That is, the utility function must capture the requirements in Section 12.6.4. Any strategy that positively contributes to this end is seen as desirable. There will be transaction costs to executing a strategy and it is thus not obvious that one should always execute a strategy, but even the awareness of an available strategy may be beneficial.

12.6.7.3 Mindset

We don't specifically propose to apply the CIRA method as such, but we do advocate to have a "Conflicting Incentives" mindset when analyzing ToS threats. That is, to keep in mind that the other principal party will have different interests and that those will govern it's actions. The consumer had better account for this and carry out actions (strategies) to mitigate or prevent negative outcomes. The key to success is to identify the appropriate utility function(s) and to identify useful strategies that address dire threats and unacceptable risks.

12.6.8 Content Ownership

A prudent question in all security is "What are the assets?" In our context, the parties will be the SP and the SC and clearly the stored "contents" is an asset. Content ownership, copyrights etc. are potential "conflicting incentives" areas. SC must, therefore, assume that he/she must contribute something to SP in the deal. One obvious contribution is information and another is potential future loyalty. Of course, actual "contents" may also be contributed. This can be a win-win situation, like a consumer uploading content to YouTube where both parties win if the content is widely shared.

The content in our case is exclusively provided by the user (SC), but may be used or licensed by the SP (depending on ToS conditions). While there may be win-win situations, we also have the distinct possibility that the equation is unbalanced and that it can even be a negative sum game [50]. The utility function may be hard to define, but in a game theoretical sense we can only assume that the function is such that at least one part will expect a positive outcome [40].

The "expect a positive outcome" part highlights the fact that there is a distinct difference between real and perceived utility. We shall not go into the psychology of perception here, but suffice to say that emotional responses are important. In [52], one discusses how emotions and limited foresight affects our perception of utility and ultimately our decision making.

12.6.9 Terms-of-Service Policy Control Options

ToS policies of most cloud service providers are presented in the form of long text in plain English that most users will never read. However, reading and understanding of such policies are crucial for security and privacy for users of these services. Since such policies are a subject to change, the continuous monitoring of security and privacy related changes is necessary and could be implemented as a part of MaaS (Monitoring-as-a-Service) [188].

One possible approach can be based on text analysis and meaning extraction with special focus on those parts of ToS policies that can potentially influence users' security and privacy. By timely identifying such threats, users could undertake necessary measures to protect of both their security and privacy as required by their own policy. Possible measures could be, for example, to enforce encryption of downloaded content to guaranty confidentiality; to enforce encryption or anonymization of some parts of the content to provide privacy; to require distribution of data among several independent service providers to increase availability, etc. It is important for users to be aware of threats that potentially may appear as result changes in ToS policies. In that, there is a need for formal representation of policies, for example in P3P style, to simplify extraction of features that can be a security and privacy threats for the users. However, since ToS policies are usually written in English we have to deal with natural language understanding. Since it is generally a difficult problem one cannot expect a perfect solution. One approach is to propose a practical approach that will help users to monitor changes in ToS but, in the end, will need human involvement to make a final decision [166].

One notable feature of many ToSs is that many of them permit the service provider to (a) unilaterally change the ToS and (b) to only inform the service user by notification on a webpage and possible by an email. Thus, even the interested consumer may not notice that there has been a change in the ToS. Even uninterpreted notification by the policy monitoring agent (PoMA) will, therefore, have value.

12.6.9.1 Case Study

Let us consider some example of ToS policies of some well-known companies. Such policies are presented in English and, therefore, a method for text analysis and meaning extraction should be developed. To illustrate our idea we have extracted sentences containing keywords *grant*, *right to use*, *license to use* or *to distribute* in ToS policies of some popular cloud service providers:

- From Linkedin: " ...you grant LinkedIn a nonexclusive, irrevocable, worldwide, perpetual, unlimited, assignable, sublicenseable, fully paid up and royalty-free *right to* us to copy, prepare derivative works of, improve, distribute, publish, remove, retain, add, process, analyze, *use* and commercialize, in any way now known or in the future discovered, any information you provide, directly or indirectly to LinkedIn, including, but not limited to, any user generated content, ideas, concepts, techniques or data to the services, you submit to LinkedIn, without any further consent, notice and/or compensation to you or to any third parties ..."
- From Instagram: " ...you hereby *grant* to Instagram a non-exclusive, fully paid and royalty-free, transferable, sub-licensable, worldwide *li-*

cense to use the Content that you post on or through the Service, subject to the Service's Privacy Policy,... "

- From Evernote: " ...you *grant* Evernote a *license to* display, perform and *distribute* your Content and to modify (for technical purposes, e.g., making sure content is viewable on smart phones as well as computers) and reproduce such Content to enable Evernote to operate the Service. "
- From Facebook: " ...you *grant* us a non-exclusive, transferable, sub-licensable, royalty-free, worldwide license *to use* any IP content that you post on or in connection with Facebook (IP License). "
- From Box: " You hereby *grant* Box and its contractors the right, *to use*, modify, adapt, reproduce, *distribute*, display and disclose Content posted on the Service solely to the extent necessary to provide the Service or as otherwise permitted by these Terms. "
- From Comcast (and Plaxo): " If you post any content to the Comcast Web Services, you hereby *grant* Comcast and its licensees a worldwide, royalty-free, non-exclusive right and *license to use*, reproduce, publicly display, publicly perform, modify, sublicense, and *distribute* the content, on or in connection with the Comcast Web Services or the promotion of the Comcast Web Services, and incorporate it in other works, in whole or in part, in any manner. "
- From Amazon: " ...you *grant* Amazon a nonexclusive, royalty-free, perpetual, irrevocable, and fully sublicensable *right to use*, reproduce, modify, adapt, publish, translate, create derivative works from, *distribute*, and display such content throughout the world in any media."

The list of keywords can be easily expanded and patterns for matching can be regular expression for example a such as: {grant | license | right} to <company_ name> or { license to {use | reproduce | modify |distribute }}.

Just reading these few extracts provides users with hints of potential threats to their security and privacy from these services, that is an understanding that service providers keep the right to use, distribute etc. their content.

12.6.9.2 Monitoring Terms-of-service Policies

In order to be aware of possible security and privacy threats, users have to be aware of changes in ToS policies. Monitoring and detection of changes should support two features: discovering of changes in ToS policies and discovering

<table>
<tr><th>ToS location</th><th>Pattern</th><th>Violation</th><th>Action</th></tr>
<tr><td>http://instagram.com/about/legal/terms/</td><td>{ grant | license | right} to { use | reproduce | modify | sublicense |distribute }</td><td>Privacy</td><td>Encryption</td></tr>
<tr><td>https://www.box.com/</td><td>-"''-</td><td>Confidentiality, Privacy</td><td>Encryption</td></tr>
<tr><td>http://www.linkedin.com/static?key=user_agreement&trk=hb_ft_userag</td><td>-"''-</td><td>Privacy</td><td>No actions available</td></tr>
</table>

Figure 12.2 Configuration table for ToS monitoring agent

of security and privacy related changes in ToS policies. The first kind of changes is easy to implement, since it means detection of any changes in a text and does not require understanding of these changes. The second feature requires at least a rudimentary understanding of texts.

Some cloud-monitoring applications have already been described in the literature, for example, SLA Violation Detection [89], Resource Usage [77], etc. However, we could not find any policy monitoring applications.

In the approach considered here, we propose to use monitoring agent running locally on user's computer and analyzing changes of ToS policy each time user use the service. The user register website's policy locations in the agent database for each cloud service he/she uses. Monitoring agent analyzes the policy text to detect changes from last time the service was used. The monitoring is based on the idea informally presented in the previous subsection. In case the change is detected, the agent will identify new sentences, removed sentences and modified sentences. By matching patterns, the agent extracts sentences (containing patterns) that may be potentially associated with security and privacy threats.

For example, pattern `license to {use | reproduce | modify | sublicense |distribute}` means that confidentiality of user's data may be violated. If it is a storage service and such feature is considered as violation of user's security policy an encryption of all data have to be activated. An example of configuration table for the PoMA is shown in Figure 12.2.

12.6.10 Cloud–Based Storage Services

12.6.10.1 File Storage and Synchronization Services

There are several publicly available cloud-based file storage services like Dropbox, Google Drive and Microsoft SkyDrive. Generally, the free services are restricted in the amount of storage they offer. These file storage services mirror a directory tree at the target computer. All files placed in the target directory will be synchronized and a copy of the files will be stored in the cloud by the service. The beauty of the scheme is that one may synchronize several computers this way.

Enterprises and businesses are understandably concerned about storing their essential business data in a public cloud storage facility. In the article "New Approaches to Security and Availability for Cloud Data" [159], the authors discuss these problems and propose some solutions. The tenants (consumers, SC) will have to be convinced that security and availability is ensured or otherwise they will tend to favour private clouds instead of public clouds. The power balance between a paying tenant (SC) and a public cloud service provider (SP) may not be balanced, but SP clearly have an incentive to accommodate the paying tenant. Our perspective is a little different from [159] in that our scope includes cases where the power balance is very different. We also assume a model where the data is mirrored at the tenants computers. Free services will also have some kind of authentication and access control, but that is very often all that is provided. Use of cryptography is clearly necessary to provide credible security for data stored in a public cloud and availability may dictate schemes similar to the RAID-inspired HAIL scheme proposed in [159]. HAIL itself is an acronym for High-Availability and Integrity Layer [46]. We do not consider a HAIL-based approach here. Instead, we do propose a model with basic security services implemented at the presentation layer. Our scheme is in concordance with Presentation layer in the Systems Interconnection (OSI) Reference Model, and this also means that the implementation is at the host (tenant) and not at the SP.

12.6.10.2 An OSI "Presentation Layer" Solution

The OSI reference model [146] is defined by the International Organization for Standardization (ISO). The OSI RM is a way to describe and characterize a communications stack in terms of abstraction functions defined at different layers. Each layer serves the layer above it and is served by the layer below. Layer 6 is the *presentation layer* and is primarily concerned with services such as data representation, encryption and decryption, converting machine dependent data to machine independent data etc. For connectionless services (IP-based), recommendation [147] applies.

12.6.10.3 The Umbrella Architecture

The *Umbrella Architecture* depicted in Figure 12.3, is proposed in [166]. It combines the services of the presentation layer security manager (PLaSM) and the policy monitoring agent (PoMA). In the scheme, a local PoMA communicates with a MaaS cloud service to carry out the actual ToS monitoring.

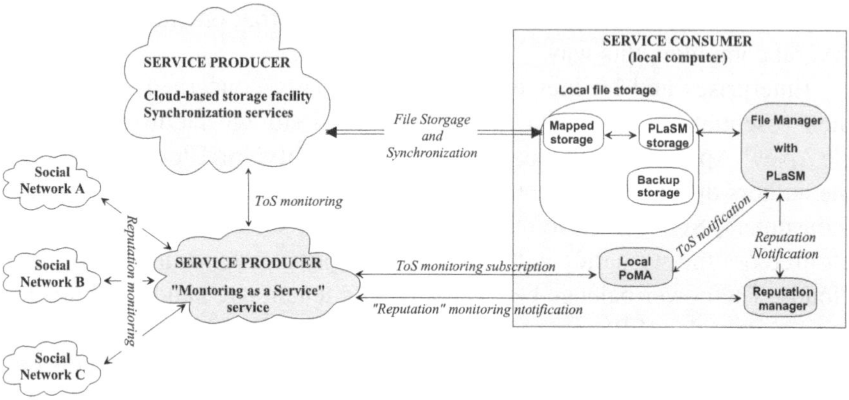

Figure 12.3 The Umbrella Architecture Concept

It also (optionally) proposes to have a *reputation manager* in the system. The reputation manager will use methods discussed in [156, 210] to extract information about the reliability and trustworthiness of the SP. The work in [210] will need to be extended to achieve full "reputation" handling.

12.6.10.4 Scope and Basic Architecture

The Umbrella architecture does not attempt to cater to all possible security services. Instead, it attempts to define a light-weight solution that may be implemented in web browsers and file browser as a simple plug-in service.

- Simple AAA Service
- Data Integrity service
- Data Confidentiality service

Services such as a "Time vault/Backup" and a "Cloud withdrawal" should also be considered.

12.6.10.5 Service Resolution

The considered model is based on a basic synchronized data storage service. It is important that the new security services fit with this model. For instance, since the synchronization is file based, the security services should also be file based. Disk encryption schemes, for instance, TrueCrypt [240] work by creating a huge encrypted file and then presenting this as a directory or a disk volume. Such a scheme would force synchronization of the entire "disk", which in most cases really is *not* what one would like.

The preliminary analysis indicates that the service resolution should be "file-based". This should include meta-information such as file names, which may be sensitive, and even files sizes. Files may also be padded up to specified block lengths, where the block length could be set according to common file system block lengths. It adopts the view that file names should be protected and that padding should be used.

12.6.10.6 The Presentation Layer Security Manager

The presentation layer security manager is a plug-in service to the "normal" file manager, and it may be available in productivity software (office packages) and in web browsers. It may be inspired by tools such as the HP ProtectTools [137] or similar.

12.6.10.7 Identification and Security Context Setup

One drawback to having a client-side solution is to have to manage security at the client-side. The PLaSM must authenticate *SC* and create a security context so as to decipher/encipher the system information and the files whenever needed. This adds complexity and its own share of security management problems, but should nevertheless be feasible. It is proposed that the SC identity is the same as used for the cloud service itself. The password, or other security credential, should for obvious reasons *not* be similar to the one used to access the cloud service. As of now, the choice of credentials and the actual security context is left for further study, but suffice to say that the credentials should be flexible in use and not themselves represent a security weakness.

12.6.10.8 Security Service Provisioning

The aim primarily is to provide data confidentiality and data integrity. This should include concealment of file/directory names and file length padding. Standard file encryption methods seem adequate here. As a minimal solution data integrity should be implemented on a per-file basis, but one may also have file system integrity built into the system. The integrity solution should not be implemented such that it itself unduely increases the synchronization activity of the cloud service.

12.7 Summary

In this chapter, we have considered implications of cloud computing on privacy. We considered intruder model and requirements that need to be satisfied to provide required level of privacy. Since previous research show that cryptographic means cannot always provide protection (especially in long term) we propose a trust-based privacy protection. We considered an approach based on subjective logic that was applied to measure/monitor level of trustworthiness of cloud service providers and explain how users have to handle their data to minimize privacy treats in the cloud.

We have also addressed problems associated with the so-called ToS threat. The ToS threat is not exclusive to cloud services, but is highlighted by the ubiquitousness and proliferation of cloud-based services. The ToS threat is mostly seen as a threat towards unpaid and publicly available services, but the threat is, in principle, generic.

Summary and Concluding Remarks

Summary

Chapter 1: Do we have a right to privacy?
This topic was explored in the first chapter. Early pioneers in the field concluded that the right "to be let alone" is a fundamental right. We looked at identifiers, various aspects of privacy and the *Privacy by Design* concepts.

Chapter 2: Legal Aspects.
This topic was explored in Chapter 2. There are many reasons why one may want privacy to be set aside for moment. Some of them are entirely to our own benefit, like allowing the emergency call center to automatically get the callers location data. We may be private citizens, but we do live in a society. Crime is a problem and for serious crimes it is customary to allow the policy to intercept our electronic communications. The justification seems to be there, but it remains that lawful interception can be subverted and what is considered lawful is only valid as long as the law enforcement agencies act according to just laws. Somewhat more mundane, but likely with more impact is the handling of digital ownership rights. Many schemes in the past have been a hazard to both security and privacy.

Chapter 3: Anonymous Communication.
What does it mean to be anonymous? How is it different from being pseudonymous? Should we allow anonymous communications? What about accountability? In Chapter 3, we took a look at MIX networks and Onion Routing systems. These systems provide different privacy services, but are generally about anonymous communications over the internet. And it's not just theory: The Tor network is for real and has many users and many onion routers all around the world.

Chapter 4: Secure Multi-party Computations (SMC).
The topic of Chapter 4 was privacy-preserving through cryptographic means.

Cryptography is the key to many of the technological privacy solutions. SMC is a special branch of cryptography with special relevance to privacy, as it provides methods for computing on encrypted data. It may sound paradoxical, but certain crypto-systems actually permit algebraic operations on encrypted data as if it were un-encrypted data.

Chapter 5: Data Mining.
Data mining in telecommunications was the topic of Chapter 5. Data mining is said to be about finding the needle in the haystack, except that there is no needle. Or in mining terms, to find a nugget of gold in a dessert of valueless sand. More seriously, we may say that it is about finding trends, associations and connections between data elements. There are a lot of buzzwords around like Big Data, but at its core data mining is about detecting patterns and correlations. Data mining is, of course, also troublesome for privacy since our digital behavior leaves a lot of data and meta-data around. If there is a pattern, the data mining tools will find it. You may use an anonymous communications channel, but unless you take care your identity will be inferred from your actions (behavior).

Chapter 6-8: Cellular Privacy.
Cellular privacy is covered in chapters 6–8. The mobile phone is the ultimate private gadget. We tend to carry it with us at all times and a lot of people don't even turn it off when they go to sleep, but keep it next to their pillow. The very nature of cellular systems makes it inevitable that the radio access network must and will know the approximate location of the mobile phones. There is even an explicit requirement (emergency calls) to be able to determine the location. Adding to this, almost all newer mobiles will have satellite position technology embedded in the device. This makes the mobile phone an excellent people tracking device. In Chapter 6 we looked at the problems and the identifiers used in the 3GPP systems (GSM, GPRS, EDGE, UMTS, LTE/LTE-Advanced) and a little bit about the access security schemes. Chapter 7 we discussed concrete technologies (GSM, UMTS and LTE) and in Chapter 8 we took a look at alternatives to the current design. Here, we demonstrated that better privacy is possible.

Chapter 9: Sensor Network.
Wireless sensor networks (WSNs) consists of spatially distributed autonomous sensors to monitor physical or environmental conditions. The

development of wireless sensor networks was largely motivated by military applications such as battlefield surveillance. Today WSNs are used in many industrial and consumer applications, including so-called eHealth contexts, but the capability for surveillance remains. Given that the WSNs are used for surveillance and data gathering purposes, there is obviously a potential for privacy invasion. We investigated various aspects such as data privacy, access privacy, location privacy, privacy aware routing and privacy-preserving data monitoring. We may conclude that while there exists solutions there seems to exist even more challenges in this field.

Chapter 10: Radio Frequency Identification (RFID).
RFID was covered in this chapter. These devices can literally be as small as a speck of dust, they have limited memory, limited processing power and limited power source. RFID is already well into the mainstream. The RFID devices have limited range and can't generally do very much, except for being a marker device. Tracking is indeed a primary feature of an RFID device, but while tracking goods may seem innocent it will also allow tracking anyone associated with the tracked item. We also came to know the esoteric *physically unclonable function* and other super light-weight attempt at security for RFID.

Chapter 11: Internet of Things (IoT).
The so-called "Internet of things" was covered in this chapter. We also used this chapter to cover trust in more detail. These devices differ from the sensor and RFID devices described in the previous chapter in that an IoT-device by definition will have internet connectivity. This changes the playground significantly and these devices are routinely exposed to the wide open internet. With respect to privacy, can you really trust devices? And how would we come to trust the devices? And last but not least, are the devices trustworthy? There can be no definitive answer, we did show how a device might act to protect our privacy. There will likely be a lot more research in this area in coming years.

Chapter 12: Privacy in the Cloud.
It has been said that there can be no privacy in the cloud. Maybe so, but clouds comes in different guises and all hope is not lost. The cloud is a buzzword like few others, but at its core we have a massively distributed data storage and execution environment. Cloud data and services are run off huge data

centers, but Virtual Machine (VM) technologies and broadband connections allow data and services to migrate. There is therefore no fixed location for a specific service, and both data and service may be located at a different physical data center the next time you use a service. This migration and lack of specific location makes it extremely hard for the user to ascertain that the data is being taken care of according to the agreement. We looked at some aspects of privacy in the cloud in this chapter, but the topic is huge and we only scratched the surface.

Concluding Remarks

> "They who can give up essential liberty to obtain a little temporary safety deserve neither liberty nor safety."
>
> – Benjamin Franklin

This books covers considerable ground, but privacy is a much bigger topic than any one book can cover. Our focus has been on technology and services that can help us with the technical aspects of privacy. And we haven't even covered field this in more than fragments. What we have left out is aspects such as personal attitude towards privacy, practical setup and configuration of internet services that actually provide privacy etc. For instance, we have covered cellular systems and the problems with identity- and location privacy in considerable depth. This is important since what leaks at the lower layers can never be remedied at the high layers. Still, in a world of smart phones and myriads of so-called Apps, it is clear that the lower layer mechanisms isn't enough either. We can prevent tracking of lower layer identifiers, but then an App may so easily leak identifiers and it may turn on location services to pinpoint the location too. Social networks will leak even more sensitive information, but this will be done at our more-or-less well-informed consent.

To really handle privacy in the future requires not only technological solutions, but also awareness, attitude and willingness to protect ourselves. Society at large does't have a fixed notion of what privacy is and should be. There is considerable plasticity and 9/11 showed this in practice. Governmental surveillance that would never have been acceptable prior to 9/11 is suddenly seen as justified.

Kids grow up with digital services from young age and they need to be taught how to act responsibly and be aware of the value of privacy. This is a tall order, but it is necessary if personal privacy is to have any meaning at all.

Bibliography

[1] 3GPP TS 04.08. Mobile radio interface layer 3 specification. 3GPP, 12 2003.
[2] 3GPP TS 21.133. 3G Security; Security Threats and Requirements. 3GPP, 12 2001.
[3] 3GPP TS 23.002. Network architecture. 3GPP, 12 2008.
[4] 3GPP TS 23.003. Numbering, addressing and identification. 3GPP, 06 2011.
[5] 3GPP TS 23.108. Mobile radio interface layer 3 specification, Core network protocols; Stage 2 . 3GPP, 06 2007.
[6] 3GPP TS 23.401. General Packet Radio Service (GPRS) enhancements for Evolved Universal Terrestrial Radio Access Network (E-UTRAN) access . 3GPP, 03 2009.
[7] 3GPP TS 29.002. Mobile Application Part (MAP) specification;. 3GPP, 03 2006.
[8] 3GPP TS 29.272. Evolved Packet System (EPS); Mobility Management Entity (MME) and Serving GPRS Support Node (SGSN) related interfaces based on Diameter protocol . 3GPP, 03 2009.
[9] 3GPP TS 31.101. UICC-terminal interface; Physical and logical characteristics. 3GPP, 01 2009.
[10] 3GPP TS 33.102. 3G Security; Security architecture. 3GPP, 09 2012.
[11] 3GPP TS 33.105. 3G Security; Cryptographic algorithm requirements. 3GPP, 06 2004.
[12] 3GPP TS 33.107. 3G Security; Lawful interception architecture and functions. 3GPP, 06 2011.
[13] 3GPP TS 33.210. 3G Security; Network Domain Security; IP network layer security. 3GPP, 06 2004.
[14] 3GPP TS 33.310. Network Domain Security (NDS); Authentication Framework (AF). 3GPP, 09 2004.
[15] 3GPP TS 33.320. Security of Home Node B (HNB) / Home evolved Node B (HeNB). 3GPP, 09 2011.
[16] 3GPP TS 33.401. 3GPP System Architecture Evolution (SAE): Security Architecture;. 3GPP, 03 2009.
[17] 3GPP TS 35.201. 3G Security; Specification of the 3GPP Confidentiality and Integrity Algorithms; Document 1: f8 and f9 Specification. 3GPP, 12 2004.
[18] 3GPP TS 35.205. 3G Security; Specification of the MILENAGE Algorithm Set: An example algorithm set for the 3GPP authentication and key generation functions f1, f1*, f2, f3, f4, f5 and f5*; Document 1: General. 3GPP, 12 2004.
[19] 3GPP TS 35.206. 3G Security; Specification of the MILENAGE Algorithm Set: An example algorithm set for the 3GPP authentication and key generation functions f1, f1*, f2, f3, f4, f5 and f5*; Document 2: Algorithm Specification. 3GPP, 12 2004.
[20] 3GPP TS 35.207. 3G Security; Specification of the MILENAGE Algorithm Set: An example algorithm set for the 3GPP authentication and key generation functions f1, f1*, f2, f3, f4, f5 and f5*; Document 3: Implementors' Test Data. 3GPP, 12 2004.

[21] 3GPP TS 35.208. 3G Security; Specification of the MILENAGE Algorithm Set: An example algorithm set for the 3GPP authentication and key generation functions f1, f1*, f2, f3, f4, f5 and f5*; Document 4: Design Conformance Test Data. 3GPP, 12 2004.

[22] 3GPP TS 35.215. Specification of the 3GPP Confidentiality and Integrity Algorithms UEA2 & UIA2; Document 1: UEA2 and UIA2 specifications. 3GPP, 06 2006. *Subject to licensing.*

[23] 3GPP TR 35.909. 3G Security; Specification of the MILENAGE Algorithm Set: An example algorithm set for the 3GPP authentication and key generation functions f1, f1*, f2, f3, f4, f5 and f5*; Document 5: Summary and results of design and evaluation. 3GPP, 12 2004.

[24] 3GPP T2 43.020. Security related network functions. 3GPP, 06 2011.

[25] 3GPP TS 55.205. Specification of the GSM-MILENAGE Algorithms: An example algorithm set for the GSM Authentication and Key Generation functions A3 and A8. 3GPP, 06 2006.

[26] B. Aboba, L. Blunk, J. Vollbrecht, J. Carlson, and H. Lewkowetz. RFC 3748: Extensible Authentication Protocol (EAP). The Internet Engineering Task Force (IETF), http://www.ietf.org/, 06 2004.

[27] Sharon Adam. Preserving authenticity in the digital age. *Library Hi Tech, ISSN: 0737-8831*, 28(4):595–604, 2010.

[28] W.H. Allen. Computer Forensics. *IEEE Security & Privacy*, (4):59–62, 08 2005.

[29] Khaled Alotaibi, Victor J. Rayward-Smith, and Beatriz de la Iglesia. Non-metric multidimensional scaling for privacy-preserving data clustering. In *Proceedings of the 12th international conference on Intelligent data engineering and automated learning*, IDEAL'11, pages 287–298, Berlin, Heidelberg, 2011. Springer-Verlag.

[30] Gary Anthes. Security in the cloud. *Commun. ACM*, 53(11):16–18, November 2010.

[31] Frederik Armknecht, Ahmad-Reza Sadeghi, Ivan Visconti, and Christian Wachsmann. *On RFID Privacy with Mutual Authentication and Tag Corruption*, volume 6123 of *LNCS*, pages 493–510. Springer, 2010.

[32] N. Asokan. Anonymity in a Mobile Computing Environment. In *Proceedings of IEEE Workshop on Mobile Computing Systems and Applications*, pages 200–204, Santa Cruz, CA, USA, 12 1994. IEEE Press.

[33] M. Atallah and W. Du. Secure multi-party computational geometry. *Algorithms and Data Structures*, pages 165–179, 2001.

[34] Giuseppe Ateniese, Amir Herzberg, Hugo Krawczyk, and Gene Tsudik. Untraceable Mobility or How to Travel Incognito. volume 31. Elsevier B.V., 4 1999.

[35] Elad Barkan, Eli Biham, and Nathan Keller. Instant Ciphertext-only Cryptanalysis of GSM Encrypted Communication. In *Proceedings of CRYPTO 2003, 23rd Annual International Cryptology Conference, Santa Barbara, CA, USA, August 17-21, 2003*, volume 2729 of *LNCS*. Springer, 2003.

[36] A. Beimel and Y. Ishai. Information-theoretic private information retrieval: A unified construction. *Automata, Languages and Programming*, pages 912–926, 2001.

[37] J. Benaloh. Dense probabilistic encryption. In *Proceedings of the Workshop on Selected Areas of Cryptography*, pages 120–128, 1994.

[38] K. Bernsmed, M.G. Jaatun, P.H. Meland, and A. Undheim. Security slas for federated cloud services. In *Availability, Reliability and Security (ARES), 2011 Sixth International Conference on*, pages 202–209, 2011.

[39] A. Bhadani, R. Shankar, and D. Vijay Rao. A computational intelligence based approach to telecom customer classification for value added services. *Advances in Intelligent Systems and Computing*, 201 AISC(VOL. 1):181–192, 2013.

[40] Ken Binmore. *Playing for real: a text on game theory*. Oxford University Press, USA, 2007.

[41] Alex Biryukov, Adi Shamir, and David Wagner. Real Time Cryptanalysis of A5/1 on a PC. In *Proceedings of Fast Software Encryption, 7th International Workshop, FSE 2000, New York, NY, USA, April 10-12, 2000*, volume 1978 of *LNCS*, pages 1 – 18. Springer, 2001.

[42] Matt Blaze, Whitfield Diffie, Ronald L Rivest, Bruce Schneier, and Tsutomu Shimomura. Minimal key lengths for symmetric ciphers to provide adequate commercial security. a report by an ad hoc group of cryptographers and computer scientists. Technical report, DTIC Document, 1996.

[43] Peter Bogetoft, Dan Lund Christensen, Ivan Damgård, Martin Geisler, Thomas Jakobsen, Mikkel Krøigaard, Janus Dam Nielsen, Jesper Buus Nielsen, Kurt Nielsen, Jakob Pagter, Michael Schwartzbach, and Tomas Toft. Secure multiparty computation goes live. In Roger Dingledine and Philippe Golle, editors, *Financial Cryptography and Data Security*, volume 5628 of *Lecture Notes in Computer Science*, pages 325–343. Springer Berlin Heidelberg, 2009.

[44] D. Boneh and M. Franklin. Identity-Based Encryption from the Weil Pairing. In *SIAM Journal of Computing*, volume 32, pages 586–615. SIAM, 2003.

[45] Dan Boneh and Matt Franklin. Identity-based encryption from the Weil pairing. In *Proceedings of the 21st Annual International Cryptology Conference on Advances in Cryptology (CRYPTO 2001)*, volume 2139 of *Springer LNCS*. Springer, 2001.

[46] Kevin D. Bowers, Ari Juels, and Alina Oprea. Hail: a high-availability and integrity layer for cloud storage. In *Proceedings of the 16th ACM conference on Computer and communications security*, CCS '09, pages 187–198, New York, NY, USA, 2009. ACM.

[47] Colin Boyd and Anish Mathuria. *Protocols for Authentication and Key Establishment*. Springer, 1 edition, 09 2003.

[48] Sandra Braman and Stephanie Roberts. Advantage isp: Terms of service as media law. *New media & society*, 5(3):422–448, 2003.

[49] Marc Briceno, Ian Goldberg, and David Wagner. GSM Cloning. Available at *http://www.isaac.cs.berkeley.edu/isaac/gsm-faq.html*, 1998.

[50] Heidi Burgess and Guy M Burgess. *Encyclopedia of conflict resolution*. Abc-Clio Santa Bárbaraˆ eCalifornia California, 1997.

[51] Jan Camenisch and Anna Lysyanskaya. A formal treatment of onion routing. In *Advances in Cryptology–CRYPTO 2005*, pages 169–187. Springer, 2005.

[52] Colin F Camerer. *Behavioral game theory: Experiments in strategic interaction*. Princeton University Press, 2011.

[53] Kim Cameron. The Laws of Identity. Microsoft Corporation; `www.identityblog.com`, 11 2005.

[54] D. Ċamilovič. Data mining and crm in telecommunications. *Serbian Journal of Management*, 3(1):61 – 72, 2008.

[55] R. Canetti, Y. Ishai, R. Kumar, M.K. Reiter, R. Rubinfeld, and R.N. Wright. Selective private function evaluation with applications to private statistics. In *Proceedings of the twentieth annual ACM symposium on Principles of distributed computing*, pages 293–304. ACM, 2001.

[56] R.S Chakraborty, S. Narasimhan, and S. Bhunia. Hardware Trojan: Threats and Emerging Solutions. In Priyank Kalla and Prabhat Mishra, editors, *Proceedings of IEEE International High Level Design Validation and Test Workshop 2009 (HLDVT'09)*, volume 14, pages 166–171. IEEE, 11 2009.

[57] David Chaum. Untraceable Electronic Mail, Return Addresses, and Digital Pseudonyms. *Communications of the ACM*, 24(2), 1981.

[58] David Chaum, Markus Jakobsson, Ronald L Rivest, Peter YA Ryan, Josh Benaloh, Miroslaw Kutylowski, and Ben Adida. *Towards Trustworthy Elections: New Directions in Electronic Voting*, volume 6000. Springer, 2010.

[59] Abbas Cheddad, Joan Condell, Kevin Curran, and Paul Mc Kevitt. Digital image steganography: Survey and analysis of current methods. *Signal Processing*, 90(3):727–752, 2010.

[60] Yao Chen and Radu Sion. On securing untrusted clouds with cryptography. In *Proceedings of the 9th annual ACM workshop on Privacy in the electronic society*, WPES '10, pages 109–114, New York, NY, USA, 2010. ACM.

[61] Benoît Chevallier-Mames, Pierre-Alain Fouque, David Pointcheval, Julien Stern, and Jacques Traoré. On some incompatible properties of voting schemes. In *Towards Trustworthy Elections*, pages 191–199. Springer, 2010.

[62] B. Chor and N. Gilboa. Computationally private information retrieval. In *Proceedings of the twenty-ninth annual ACM symposium on Theory of computing*, pages 304–313. ACM, 1997.

[63] B. Chor, O. Goldreich, E. Kushilevitz, and M. Sudan. Private information retrieval. In *Foundations of Computer Science, 1995. Proceedings., 36th Annual Symposium on*, pages 41–50. IEEE, 1995.

[64] Kazimierz Choros. Real anomaly detection in telecommunication multidimensional data using data mining techniques. In Jeng-Shyang Pan, Shyi-Ming Chen, and NgocThanh Nguyen, editors, *Computational Collective Intelligence. Technologies and Applications*, volume 6421 of *Lecture Notes in Computer Science*, pages 11–19. Springer Berlin Heidelberg, 2010.

[65] Chi-Yin Chow, Mohamed F. Mokbel, and Tian He. A privacy-preserving location monitoring system for wireless sensor networks. *IEEE Transactions on Mobile Computing*, 10(1):94–107, 2011.

[66] Mihai Christodorescu, Reiner Sailer, Douglas Lee Schales, Daniele Sgandurra, and Diego Zamboni. Cloud security is not (just) virtualization security: a short paper. In *Proceedings of the 2009 ACM workshop on Cloud computing security*, CCSW '09, pages 97–102, New York, NY, USA, 2009. ACM.

[67] V. Ciriani, S. Capitani di Vimercati, S. Foresti, and P. Samarati. k-anonymity. In Ting Yu and Sushil Jajodia, editors, *Secure Data Management in Decentralized Systems*, volume 33 of *Advances in Information Security*, pages 323–353. Springer US, 2007.

[68] Chris Clifton, Murat Kantarcioglu, Jaideep Vaidya, Xiaodong Lin, and Michael Y. Zhu. Tools for privacy preserving distributed data mining. *SIGKDD Explor. Newsl.*, 4(2):28–34, December 2002.

[69] Data Privacy & Integrity Advisory Committee. Report No.2006-02; The Use of RFID for Human Identity Verification (Adopted December 6, 2006). Technical report, Department of Homeland Security, 12 2006.

[70] Data Privacy & Integrity Advisory Committee. The use of RFID for Human Identification, A draft report from DHS Emerging Applications and Technology subcommittee to the Full Data Privacy and Integrity Advisory Committee, Version 1.0. Technical report, Department of Homeland Security, 2006.

[71] Jean-Sebastien Coron, David Naccache, and Paul Kocher. Statistics and secret leakage. *ACM Trans. Embed. Comput. Syst.*, 3(3):492–508, August 2004.

[72] George Danezis and Claudia Diaz. A survey of anonymous communication channels. *Computer Communications*, 33, 2008.

[73] George Danezis, Claudia Diaz, and Paul Syverson. Systems for anonymous communication. *Handbook of Financial Cryptography and Security, Cryptography and Network Security Series*, pages 341–389, 2009.

[74] Paolo D'Arco, Alessandra Scafuro, and Ivan Visconti. *Revisiting DoS Attacks and Privacy in RFID-Enabled Networks*, volume 5804 of *LNCS*, pages 76–87. Springer, 2009.

[75] Jing Deng, Richard Han, and Shivakant Mishra. Decorrelating wireless sensor network traffic to inhibit traffic analysis attacks. *Pervasive Mob. Comput.*, 2(2):159–186, April 2006.

[76] Robert H. Deng, Yingjiu Li, Moti Yung, and Yunlei Zhao. *A New Framework for RFID Privacy*, volume 6345 of *LNCS*, pages 1–18. Springer, 2010.

[77] M. Dhingra, J. Lakshmi, and S. K. Nandy. Resource usage monitoring in clouds. In *Grid Computing (GRID), 2012 ACM/IEEE 13th International Conference on*, pages 184–191, 2012.

[78] T. Dierks and E. Rescorla. RFC 5246: The Transport Layer Security (TLS) Protocol Version 1.2. The Internet Engineering Task Force (IETF), http://www.ietf.org/, 08 2008.

[79] Roger Dingledine, Nick Mathewson, and Paul Syverson. Tor: The Second-Generation Onion Router. In *Proceedings of the 13th conference on USENIX Security Symposium (SSYM'04)*, volume 13. USENIX Association, 2004.

[80] Roger Dingledine, Nick Mathewson, and Paul Syverson. Deploying low-latency anonymity: Design challenges and social factors. *Security & Privacy, IEEE*, 5(5):83–87, 2007.

[81] Danny Dolev and Andrew C. Yao. On the Security of Public-Key Protocols. *IEEE Transactions on Information Theory*, 29(2):198–208, 3 1983.

[82] Julie Doyle, Herna Viktor, and Eric Paquet. Long-term digital preservation: preserving authenticity and usability of 3-D data. *International Journal on Digital Libraries*, 10(1):33–47, 2010.

[83] W. Du and M.J. Atallah. Privacy-preserving cooperative scientific computations. In *14th IEEE Computer Security Foundations Workshop*, pages 273–282. Citeseer, 2001.

[84] W. Du and M.J. Atallah. Secure multi-party computation problems and their applications: a review and open problems. In *Proceedings of the 2001 workshop on New security paradigms*, pages 13–22. ACM, 2001.

[85] W. Du and Z. Zhan. A practical approach to solve secure multi-party computation problems. In *Proceedings of the 2002 workshop on New security paradigms*, pages 127–135. ACM, 2002.

[86] Orr Dunkelman, Nathan Keller, and Adi Shamir. A practical-time attack on the a5/3 cryptosystem used in third generation gsm telephony. Cryptology ePrint Archive, Report 2010/013, 2010. http://eprint.iacr.org/.

[87] Peter Eckersley. How Unique Is Your Web Browser? Electronic Frontier Foundation, panopticlick.eff.org/browser-uniqueness.pdf.

[88] Taher ElGamal. A public key cryptosystem and a signature scheme based on discrete logarithms. In *Advances in Cryptology*, pages 10–18. Springer, 1985.

[89] V.C. Emeakaroha, T.C. Ferreto, M.A.S. Netto, I. Brandic, and C.A.F. De Rose. Casvid: Application level monitoring for sla violation detection in clouds. In *Computer Software and Applications Conference (COMPSAC), 2012 IEEE 36th Annual*, pages 499–508, July.

[90] K. Traub et al. The EPCglobal Architecture Framework. Technical Report Ver.1.4, GS1, Brussels, Belgium, 12 2010.

[91] Nigel Smart et al. ECRYPT II Yearly Report on Algorithms and Keysizes (2011-2012). Report D.SPA.20, ECRYPT II (ICT-2007-216676), 09 2012.

[92] ETSI. ES 201 158 Telecommunications security; Lawful Interception (LI); Requirements for network functions. ETSI, 2003.

[93] ETSI. SR 002 180 Requirements for communication of citizens with authorities/organizations in case of distress (emergency call handling). ETSI, 12 2003.

[94] ETSI. TR 101 943 Lawful Interception (LI); Concepts of Interception in a Generic Network Architecture. ETSI, 2006.

[95] ETSI. ES 201 671 Lawful Interception (LI); Handover Interface for the Lawful Interception of Telecommunications Traffic. ETSI, 2007.

[96] ETSI. TR 102 180 Emergency Communications (EMTEL); Basis of requirements for communication of individuals with authorities/organizations in case of distress (Emergency call handling) Technical Report. ETSI, 09 2011.

[97] ETSI. TR 102 299 Emergency Communications (EMTEL); Collection of European Regulatory Texts and orientations. ETSI, 08 2011.

[98] ETSI. TS 101 671 Lawful Interception (LI); Handover interface for the lawful interception of telecommunications traffic. ETSI, 2011.

[99] European Council. European Council Resolution January 1995 JAI 42 Rev 28197/2/95 (Official Journal reference 96C 329/01 4 November 1996). Resolution, EU, 1995.

[100] European Council. DECLARATION ON COMBATING TERRORISM (Brussels, 25 March 2004). Technical report, EU, 2004.

[101] European Parliament/European Council. DIRECTIVE 2002/58/EC OF THE EUROPEAN PARLIAMENT AND OF THE COUNCIL of 12 July 2002 concerning the processing of personal data and the protection of privacy in the electronic communications sector (Directive on privacy and electronic communications). Directive 58/EC, EU, 2002.

[102] European Parliament/European Council. DIRECTIVE 2006/24/EC OF THE EUROPEAN PARLIAMENT AND OF THE COUNCIL of 15 March 2006 on the retention of data generated or processed in connection with the provision of publicly available electronic communications services or of public communications networks and amending Directive 2002/58/EC, Available at http://www.statewatch.org/news/2006/feb/st03677-05.pdf. Directive 24/EC, EU, 2006.

[103] Nathan S Evans, Roger Dingledine, and Christian Grothoff. A practical congestion attack on tor using long paths. In *Proceedings of the 18th USENIX Security Symposium*, pages 33–50, 2009.

[104] S. Even, O. Goldreich, and A. Lempel. A randomized protocol for signing contracts. *Communications of the ACM*, 28(6):637–647, 1985.

[105] Alexandre Evfimievski, Ramakrishnan Srikant, Rakesh Agarwal, and Johannes Gehrke. Privacy preserving mining of association rules. *Inf. Syst.*, 29(4):343–364, June 2004.

[106] Nicolas Falliere, Liam O Murchu, and Eric Chien. Symantec: W32.Stuxnet Dossier (v1.4). `www.symantec.com/connect/blogs/w32stuxnet-dossier`, 01 2011.

[107] H. Federrath, A. Jerichow, and A. Pfitzmann. MIXes in Mobile Communication Systems: Location Management with Privacy. In *Proceedings of the First Intern. Workshop on Information Hiding*, volume 1174 of *Lecture Notes in Computer Science*, pages 121–135, Cambridge, UK, 05 1996. Springer.

[108] Joan Feigenbaum, Aaron Johnson, and Paul Syverson. Preventing active timing attacks in low-latency anonymous communication. In *Privacy Enhancing Technologies*, pages 166–183. Springer, 2010.

[109] Joan Feigenbaum, Aaron Johnson, and Paul Syverson. Probabilistic analysis of onion routing in a black-box model. *ACM Transactions on Information and System Security (TISSEC)*, 15(3):14, 2012.

[110] Dinei Florêncio and Cormac Herley. Where Do All The Attacks Go? Microsoft Research; Technical Report MSR-TR-2011-74, 06 2011.

[111] Bob Fox, Nick Gurney, and Rob van den Dam. The natural fit of cloud with telecommunications: Winning in a new game through new business models, 2012.

[112] Benjamin Franklin. The quote is part of his notes for a proposition at the pennsylvania assembly. In *Memoirs of the life and writings of Benjamin Franklin*,
Written by Benjamin Franklin and William T. Franklin. Original dated to 1818 and printed in London; now digitalized and made available by Google Books., 1818.

[113] M. Freedman, K. Nissim, and B. Pinkas. Efficient private matching and set intersection. In *Advances in Cryptology-EUROCRYPT 2004*, pages 1–19. Springer, 2004.

[114] K. Frikken, M. Atallah, and C. Zhang. Privacy-preserving credit checking. In *Proceedings of the 6th ACM conference on Electronic commerce*, pages 147–154. ACM, 2005.

[115] K.B. Frikken and M.J. Atallah. Privacy preserving electronic surveillance. In *Proceedings of the 2003 ACM workshop on Privacy in the electronic society*, pages 45–52. ACM, 2003.

[116] Benjamin C. M. Fung, Ke Wang, Rui Chen, and Philip S. Yu. Privacy-preserving data publishing: A survey of recent developments. *ACM Comput. Surv.*, 42(4):14:1–14:53, June 2010.

[117] S.L. Garfinkel. Digital forensics reseach: The next 10 years. *Digital Investigation*, 7(Supplement 1):S64–S73, 08 2010.

[118] S.L. Garfinkel, A. Juels, and R. Pappu. RFID Privacy: An Overview of Problems and Proposed Solutions. *IEEE Computer & Privacy Magazine*, 3(3):34–43, 05-06 2005.

[119] Craig Gentry. Computing arbitrary functions of encrypted data. *Commun. ACM*, 53(3):97–105, March 2010.

[120] Y. Gertner, Y. Ishai, E. Kushilevitz, and T. Malkin. Protecting data privacy in private information retrieval schemes. In *Proceedings of the thirtieth annual ACM symposium on Theory of computing*, pages 151–160. ACM, 1998.

[121] Jaeseung Go and Kwangjo Kim. Wireless Authentication Protocol Preserving User Anonymity. In *The 2001 Symposium on Cryptography and Information Security (SCIS 2001)*, pages 26–36, Oiso, Japan, 1 2001. IEICE.

[122] O. Goldreich. Foundations of cryptography. *fragments of a book*, 2001.

[123] S. Goldwasser. Multi party computations: past and present. In *Proceedings of the sixteenth annual ACM symposium on Principles of distributed computing*, pages 1–6. ACM, 1997.

[124] Dieter Gollmann. Analysing security protocols. In *Formal Aspects of Security*, pages 71–80. Springer, 2003.

[125] Dan Goodin. Does your smartphone run Carrier IQ? The Register, `www.theregister.co.uk/2011/12/01/apple_sprint_carrier_iq/`, 12 2011.

[126] Tyrone Grandison and Morris Sloman. A survey of trust in internet applications. *Communications Surveys & Tutorials, IEEE*, 3(4):2–16, 2000.

[127] T. Greene. Former nsa tech chief: I dont trust the cloud. *Network World*, 2010.

[128] Jenny Gustavsson, Christel Cederberg, Ulf Sonesson, Robert van Otterdijk, and Alexandre Meybeck. Global food losses and food waste. Technical report, Food and Agriculture Organization of the United Nations (FAO) / Swedish Institute for Food and Biotechnology (SIK), Rome, Italy, 05 2011.

[129] J. Han, M. Kamber, and J. Pei. *Data Mining: Concepts and Techniques*. The Morgan Kaufmann, 3rd edition, 2011.

[130] ShuiHua Han, S. Leung, and Zongwei Luo. Tamper Detection in the EPC Network Using Digital Watermarking. *IEEE Computer & Privacy Magazine*, 9(5):62–69, 09-10 2011.

[131] C. He and J. C. Mitchell. Security Analysis and Authentication Improvement for IEEE 802.11i Specification. In *Proceedings of the 12th Annual Network and Distributed System Security Symposium (NDSS'05)*. The Internet Society 2005, 2005.

[132] Jens Hermans, Andreas Pashalidis, Frederik Vercauteren, and Bart Preneel. *A New RFID Privacy Model*, volume 6879 of *LNCS*, pages 568–587. Springer, 2011.

[133] Nicholas Hopper, Eugene Y Vasserman, and Eric Chan-Tin. How much anonymity does network latency leak? In *Proceedings of the 14th ACM conference on Computer and communications security*, pages 82–91. ACM, 2007.

[134] Günther Horn, Dan Forsberg, Wolf-Dietrich Moeller, and Valtteri Niemi. *LTE Security*. Wiley, England, 2010.

[135] Tim Hornyak. RFID Powder. Scientific American, 2 2008.

[136] R. Housley, W. Polk, W. Ford, and D. Solo. RFC 3280: Internet X.509 Public Key Infrastructure Certificate and Certificate Revocation List (CRL) Profile. The Internet Engineering Task Force (IETF), http://www.ietf.org/, 04 2002.

[137] HP. HP ProtectTools Security Software; technical white paper, 2010.

[138] Shin-Yuan Hung, David C. Yen, and Hsiu-Yu Wang. Applying data mining to telecom churn management. *Expert Systems with Applications*, 31(3):515 – 524, 2006.

[139] IEEE 802:11i. IEEE Standard for Information Technology; Telecommunications and Information Exchange Between Systems; Local and Metropolitan Area Networks; Specific Requirements Part 11: Wireless LAN Medium Access Control (MAC) and

Physical Layer (PHY) Specifications Amendment 6: Medium Access Control (MAC) Security Enhancements. IEEE, 2004.

[140] Geir M. Køien and Thomas Haslestad. Security Aspects of 3G-WLAN Interworking. *IEEE Communications magazine*, 41(11):82–89, 11 2002.

[141] ISO/IEC. Identification cards - Contactless integrated circuit(s) cards - Vicinity cards - Part 2: Air interface and initialization. Standard 15693-2, ISO/IEC, Geneva, Switzerland, 12 2006.

[142] ISO/IEC. Information technology – Radio frequency identification for item management – Part 1: Reference architecture and definition of parameters to be standardized. Standard 18000-1, ISO/IEC, Geneva, Switzerland, 06 2008.

[143] ISO/IEC. Identification cards – Contactless integrated circuit cards – Vicinity cards – Part 3: Anticollision and transmission protocol. Standard 15693-3, ISO/IEC, Geneva, Switzerland, 04 2009.

[144] ISO/IEC. Identification cards - Contactless integrated circuit(s) cards - Vicinity cards - Part 1: Physical characteristics. Standard 15693-1, ISO/IEC, Geneva, Switzerland, 9 2010.

[145] ISO/IEC. Information technology – Radio frequency identification (RFID) for item management – Software system infrastructure – Part 1: Architecture. Standard 24791-1, ISO/IEC, Geneva, Switzerland, 08 2010.

[146] ISO/IEC 7498-1. Information technology – Open Systems Interconnection – Basic Reference Model: The Basic Model. In *ISO/IEC 7498-1:1994*. ISO, Geneva, Switzerland, 1994.

[147] ISO/IEC 9576-1. Information technology – Open Systems Interconnection – Connectionless Presentation protocol: Protocol specification. In *ISO/IEC 7498-1:1994*. ISO, Geneva, Switzerland, 1995.

[148] ITU-T. Rec. E.164; The international public telecommunication numbering plan. ITU, 02 2005.

[149] ITU-T. Rec. E.212; The international identification plan for public networks and subscriptions. ITU, 5 2008.

[150] Markus Jakobsson, Ari Juels, and Ronald L Rivest. Making mix nets robust for electronic voting by randomized partial checking. In *Proceedings of the 11th USENIX Security Symposium*, pages 339–353, 2002.

[151] Ying Jian, Shigang Chen, Zhan Zhang, and Liang Zhang. Protecting receiver-location privacy in wireless sensor networks. In *INFOCOM 2007. 26th IEEE International Conference on Computer Communications. IEEE*, pages 1955–1963, 2007.

[152] Y. Jin, N. Kupp, and Y. Makris. Experiences in Hardware Trojan Design and Implementation. In *Proceedings of IEEE International Workshop on Hardware Oriented Security and Trust (HOST'09)*, pages 50–57, 08 2009.

[153] Aaron M Johnson, Paul Syverson, Roger Dingledine, and Nick Mathewson. Trust-based anonymous communication: Adversary models and routing algorithms. In *Proceedings of the 18th ACM conference on Computer and communications security*, pages 175–186. ACM, 2011.

[154] Audun Jøsang. An algebra for assessing trust in certification chains. In *Proceedings of the Network and Distributed Systems Security Symposium (NDSS'99). The Internet Society*, 1999.

[155] Audun Jøsang. A logic for uncertain probabilities. *Int. J. Uncertain. Fuzziness Knowl.-Based Syst.*, 9(3):279–311, June 2001.

[156] Audun Jøsang. Subjective logic. *CA: University of Oslo*, 2010.

[157] Audun Jøsang, Roslan Ismail, and Colin Boyd. A survey of trust and reputation systems for online service provision. *Decision support systems*, 43(2):618–644, 2007.

[158] Antoine Joux. A One Round Protocol for Tripartite DiffieHellman. In Wieb Bosma, editor, *Proceedings ANTS-IV 2000*, volume 1838 of *Lecture Notes in Computer Science*, pages 385 – 394, Leiden, The Netherlands, 07 2000. Springer.

[159] Ari Juels and Alina Oprea. New approaches to security and availability for cloud data. *Commun. ACM*, 56(2):64–73, February 2013.

[160] Pandurang Kamat, Y. Zhang, W. Trappe, and C. Ozturk. Enhancing source-location privacy in sensor network routing. In *Distributed Computing Systems, 2005. ICDCS 2005. Proceedings. 25th IEEE International Conference on*, pages 599–608, 2005.

[161] M. Kantarcioglu and C. Clifton. Privacy-preserving distributed mining of association rules on horizontally partitioned data. *Knowledge and Data Engineering, IEEE Transactions on*, 16(9):1026–1037, 2004.

[162] Douglas J Kelly, Richard A Raines, Michael R Grimaila, Rusty O Baldwin, and Barry E Mullins. A survey of state-of-the-art in anonymity metrics. In *Proceedings of the 1st ACM workshop on Network Data Anonymization*, pages 31–40. ACM, 2008.

[163] S. Kent and K. Seo. RFC 4301: Security Architecture for the Internet Protocol. The Internet Engineering Task Force (IETF), http://www.ietf.org/, 12 2005.

[164] D. Kesdogan, H. Federrath, A. Jerichow, and A. Pfitzmann. Location Management Strategies Increasing Privacy in Mobile Communication. In *Proceedings of IFIP SEC 1996*, pages 39–48, 1996.

[165] J. Kilian. Founding crytpography on oblivious transfer. In *Proceedings of the twentieth annual ACM symposium on Theory of computing*, pages 20–31. ACM, 1988.

[166] Geir Køien and Vladimir Oleshchuk. Addressing the terms-of-service threat; client-side security and policy control for free file storage services, 2013.

[167] Geir M. Køien. An Introduction to Access Security in UMTS. *IEEE Wireless Communications magazine*, 11(1):8–18, 2 2004.

[168] Geir M. Køien. Principles for Cellular Access Security. In Sanna Liitmatainen and Teemupekka Virtanen, editors, *Proceedings of NORDSEC 2004*, pages 65–72, Espoo, Finland, 11 2004. NORDSEC, HUT, Finland.

[169] Geir M. Køien. Privacy Enhanced Cellular Access Security. In *Proceedings of the 2005 ACM Workshop on Wireless Security*, pages 57–66, Cologne, Germany, 9 2005. ACM SIGmobile, ACM Press. *The paper received the Best Paper Award.*

[170] Geir M. Køien. Privacy Enhanced Mobile Authentication. *Wireless Personal Communications*, 40(3):443–455, 02 2007. doi: 10.1007/s11277-006-9202-y.

[171] Geir M. Køien. RFID and Privacy. *Telektronikk*, 103(2):77–83, 2007.

[172] Geir M. Køien. Subscriber Privacy in Cellular Systems. *Telektronikk*, 103(2):39–51, 2007.

[173] Geir M. Køien. *Entity Authentication and Personal Privacy in Future Cellular Systems.* (River Publishers' Series in Standardisation), Aalborg, Denmark, 2009.

[174] Geir M. Køien. Access Security in 3GPP-based Mobile Broadband Systems. *Telektronikk*, 106(1):97–104, 2010.

[175] Geir M. Køien. Privacy Enhanced Device Access. In Ramjee Prasad, editor, *Proceedings of 3rd International ICST Conference on Security and Privacy in Mobile Information and Communication Systems (MobiSec 2011)*, volume 3, 2011.

[176] Geir M. Køien. Reflections on Trust in Devices: An Informal Survey of Human Trust in an Internet-of-Things Context. *Wireless Personal Communications (Springer)*, 61(3):495–510, 12 2011.

[177] Geir M. Køien and Vladimir A. Oleshchuk. Privacy-Preserving Spatially Aware Authentication Protocols; Analysis and Solution. In Svein J. Knapskog, editor, *Proceedings of NORDSEC 2003*, pages 161–174, Gjøvik, Norway, 10 2003. NORDSEC, NTNU, Norway.

[178] Geir M. Køien and Vladimir A. Oleshchuk. Spatio-Temporal Exposure Control. In *Proceedings of IEEE PIMRC 2003*, volume 14, pages 2760–2764, Beijing, China, 9 2003. IEEE PIMRC, IEEE Press.

[179] Geir M. Køien and Vladimir A. Oleshchuk. Location Privacy for Cellular Systems; Analysis and Solution. In George Danezis and David Martin, editors, *Proceedings of Privacy Enhancing Technologies workshop (PET 2005)*, volume 3856 of *Lecture Notes in Computer Science*, pages 40–58, Cavtat, Croatia, 5 2005. Springer.

[180] G.M. Køien and V.A. Oleshchuk. Personal privacy in a digital world. *Telektronikk*, 103:4–19, 2007.

[181] Karl Koscher, Ari Juels, Vjekoslav Brajkovic, and Tadayoshi Kohno. EPC RFID tag security weaknesses and defenses: passport cards, enhanced drivers licenses, and beyond. In *Proceedings of the 16th ACM conference on Computer and Communications Security (CCS'09)*, pages 33–42. ACM Press, 11 2009.

[182] Steve Kremer, Mark Ryan, and Ben Smyth. *Election verifiability in electronic voting protocols*. Springer, 2010.

[183] Marc Langheinrich. A survey of RFID privacy approaches. *Personal and Ubiquitous Computing*, 13(6), 8 2009.

[184] Bin Li, Junhui He, Jiwu Huang, and Yun Qing Shi. A survey on image steganography and steganalysis. *Journal of Information Hiding and Multimedia Signal Processing*, 2(2):142–172, 2011.

[185] Raymond A. Lorie. Long term preservation of digital information. In Edward A. Fox and Christine L. Borgman, editors, *Proceedings of the 1st ACM/IEEE-CS joint conference on Digital libraries (JCDL '01)*, pages 346–352. ACM/IEEE, 2001.

[186] Changshe Ma, Yingjiu Li, Robert H. Deng, and Tieyan Li. RFID Privacy: Relation Between Two Notions, Minimal Condition, and Efficient Construction. In *Proceedings of the 16th ACM conference on Computer and Communications Security (CCS'09)*, pages 54–65. ACM Press, 11 2009.

[187] Alfred J. Menezes, Paul C. van Oorschot, and Scott A. Vanstone. *Handbook of Applied Cryptography (Revised Reprint with Updates (5th printing))*. CRC Press, Boca Raton, Florida, USA, 2001.

[188] S. Meng and L. Liu. Enhanced monitoring-as-a-service for effective cloud management, 2012.

[189] Aikaterini Mitrokotsa, Melanie R. Rieback, and Andrew S. Tanenbaum. Classifying RFID attacks and defenses. *Information Systems Frontiers*, 12(5):491–505, 11 2010.

[190] P.A. Moskowitz, R.J. von Gutfeld, and Guenter Karjoth. System and method for disabling RFID tags. US Patent 7,737,853 B2, 06 2010.

[191] D. Naccache and J. Stern. A new public key cryptosystem based on higher residues. In *Proceedings of the 5th ACM conference on Computer and communications security*, pages 59–66. ACM, 1998.

[192] M. Naor and B. Pinkas. Efficient oblivious transfer protocols. In *Proceedings of the twelfth annual ACM-SIAM symposium on Discrete algorithms*, pages 448–457. Society for Industrial and Applied Mathematics, 2001.

[193] G. Narayanaswamy, S.K. Jagannatha, and D.W. Engels. Blocking Reader: Design and implementation of a low-cost passive UHF RFID Blocking Reader. In D.W. Engels, editor, *Proceedings of the 2010 IEEE International Conference on RFID*. IEEE Press, 04 2010.

[194] Ching Yu Ng, Willy Susilo, Yi Mu, and Rei Safavi-Naini. *RFID Privacy Models Revisited*, volume 5283 of *LNCS*, pages 251–266. Springer, 2008.

[195] Ching Yu Ng, Willy Susilo, Yi Mu, and Rei Safavi-Naini. *New Privacy Results on Synchronized RFID Authentication Protocols against Tag Tracing*, volume 5789 of *LNCS*, pages 321–336. Springer, 2009.

[196] Valtteri Niemi and Kaisa Nyberg. *UMTS Security*. Wiley, England, 2003.

[197] NIST. Federal Information Processing Standards Publication 197; ADVANCED ENCRYPTION STANDARD (AES). National Institute of Standards and Technology (NIST), 11 2001.

[198] NIST. NIST Special Publication 800-38A; 2001 Edition; Recommendation for Block Cipher Modes of Operation. National Institute of Standards and Technology (NIST), 12 2001.

[199] John Oates. Microsoft admits WGA update phones home. The Register, `www.theregister.co.uk/2007/03/09/ms_wga_phones_home/`, 03 2007.

[200] Office of the Information and Privacy Commissioner of Ontario. "Privacy by Design: The 7 Foundational Principles", In *Privacy by Design: Time to Take Control*. `www.privacybydesign.ca`, Ontario, Canada, 01 2011.

[201] V. Oleshchuk and V. Zadorozhny. Trust-aware query processing in data intensive sensor networks. In *Sensor Technologies and Applications, 2007. SensorComm 2007. International Conference on*, pages 176–180, 2007.

[202] Vladimir A Oleshchuk and Geir M. Køien. Security and privacy in the cloud a long-term view. In *Wireless Communication, Vehicular Technology, Information Theory and Aerospace & Electronic Systems Technology (Wireless VITAE), 2011 2nd International Conference on*, pages 1–5. IEEE, 2011.

[203] VladimirA. Oleshchuk. Privacy preserving monitoring and surveillance in sensor networks. In Parimala Thulasiraman, Xubin He, TonyLi Xu, MiesoK. Denko, RuppaK. Thulasiram, and LaurenceT. Yang, editors, *Frontiers of High Performance Computing and Networking ISPA 2007 Workshops*, volume 4743 of *Lecture Notes in Computer Science*, pages 485–492. Springer Berlin Heidelberg, 2007.

[204] Stanley RM Oliveira and Osmar R Zaiane. Privacy preserving clustering by data transformation. In *Proceedings of the 18th Brazilian symposium on databases*, pages 304–318. Huazhong University of Science and Technology Press, 2003.

[205] Lasse Øverlier and Paul Syverson. Location Hidden Services and Valet Nodes. *Telektronikk*, 103(2):52–60, 2007.

[206] P. Paillier. Public-key cryptosystems based on composite degree residuosity classes. In *Advances in CryptologyEUROCRYPT99*, pages 223–238. Springer, 1999.

[207] Radu-Ioan Paise and Serge Vaudenay. Mutual Authentication in RFID. In *Proceedings of the 2008 ACM symposium on Information, computer and communications security (ASIACCS'08)*, pages 292–299. ACM Press, 03 2008.

[208] Choonsik Park, Kazutomo Itoh, and Kaoru Kurosawa. Efficient anonymous channel and all/nothing election scheme. In *Advances in CryptologyEUROCRYPT93*, pages 248–259. Springer, 1994.

[209] Dusko Pavlovic. Gaming security by obscurity. In *Proceedings of New Security Paradigms Workshop (NSPW 2011)*, page 15 pp. ACM, 09 2011.

[210] Konstantinos Pelechrinis, Vladimir Zadorozhny, and Vladimir Oleshchuk. Automatic evaluation of information provider reliability and expertise. *SIS-2011-04-TELE-001-Technical report*, 2011.

[211] Andreas Pfitzmann and Marit Hansen. A terminology for talking about privacy by data minimization: Anonymity, Unlinkability, Undetectability, Unobservability, Pseudonymity, and Identity Management. Technical Report Ver 0.34, TU Dresden, 08 2010.

[212] Benny Pinkas. Cryptographic techniques for privacy-preserving data mining. *SIGKDD Explor. Newsl.*, 4(2):12–19, December 2002.

[213] V. Prevelakis and D. Spinellis. The athens affair. *IEEE Spectrum*, 44(7):26 – 33, 7 2007.

[214] M.O. Rabin. How to exchange secrets by oblivious transfer. Technical report, Technical Report TR-81, Harvard Aiken Computation Laboratory, 1981.

[215] Lisa Rajbhandari and Einar Snekkenes. Intended actions: Risk is conflicting incentives. In Dieter Gollmann and FelixC. Freiling, editors, *Information Security*, volume 7483 of *Lecture Notes in Computer Science*, pages 370–386. Springer Berlin Heidelberg, 2012.

[216] Mark Rasch. Windows genuine disadvantage; A step too far? The Register, `www.theregister.co.uk/2006/07/07/wga_disadvantage/`, 07 2006.

[217] Thomas Ristenpart, Eran Tromer, Hovav Shacham, and Stefan Savage. Hey, you, get off of my cloud: exploring information leakage in third-party compute clouds. In *Proceedings of the 16th ACM conference on Computer and communications security*, CCS '09, pages 199–212, New York, NY, USA, 2009. ACM.

[218] Greg Rose and Geir M. Køien. Access Security in CDMA200, Including a Comparison with UMTS Access Security. *IEEE Wireless Communications magazine*, 11(1):19–25, 2 2004.

[219] Mark Russinovich. Sony, Rootkits and Digital Rights Management Gone Too Far. TechNet Blogs, `blogs.technet.com/b/markrussinovich/archive/2005/10/31/sony-rootkits-and-digital-rights-management-gone-too-far.aspx`, 10 2005.

[220] Ahmad-Reza Sadeghi, Ivan Visconti, and Christian Wachsmann. *Enhancing RFID Security and Privacy by Physically Unclonable Functions*, pages 3–37. Towards Hardware-Intrinsic Security. Springer, 2010.

[221] Didier Samfat, Refik Molva, and N. Asokan. Untraceability in Mobile Networks. In *The First International Conference on Mobile Computing and Networking (ACM MOBICOM 95)*, pages 26–36, Berkely, California, USA, 11 1995. ACM Press.

[222] K. Sampigethaya and R. Poovendran. A survey on mix networks and their secure applications. *Proceedings of the IEEE*, 94(12):2142–2181, 2006.

[223] Bruce Schneier. *Beyond Fear: Thinking Sensibly about Security in an Uncertain World.* Copernicus Books, 2003.

[224] Bruce Schneier. Real Story of the Rogue Rootkit. Wired Magazine, `www.wired.com/politics/security/commentary/securitymatters/2005/11/69601`, 11 2005.

[225] Bruce Schneier. *'Liars & Outliers; Enabling the Trust That Society Needs to Thrive.* Wiley, 2012.

[226] Bruce Schneier. Terms of service as a security threat, 2013.

[227] A. Shamir. Identity-based cryptosystems and signature schemes. In *Proc. CRYPTO 1984 (LNCS 196)*, LNCS, pages 47–54. Springer, 1984.

[228] Sarah Spiekermann and Sergei Evdokimov. Critical RFID Privacy-Enhancing Technologies. *IEEE Computer & Privacy Magazine*, 7(2):56–62, 03-04 2009.

[229] S. Subashini and V. Kavitha. A survey on security issues in service delivery models of cloud computing. *Journal of Network and Computer Applications*, 34(1):1 – 11, 2011.

[230] Berit Svendsen. The EU Directive on Data Retention - An End to Justify the Means. *Telektronikk*, 103(2):31–32, 2007.

[231] L. Sweeney. Achieving k-anonymity privacy protection using generalization and suppression. *International Journal of Uncertainty Fuzziness and Knowledge-Based Systems*, 10(5):571–588, 2002.

[232] Paul Syverson. A peel of onion. In *Proceedings of the 27th Annual Computer Security Applications Conference*, pages 123–137. ACM, 2011.

[233] Paul Syverson. Why im not an entropist. In Bruce Christianson, James A. Malcolm, Vashek Maty, and Michael Roe, editors, *Security Protocols XVII*, volume 7028 of *Lecture Notes in Computer Science*, pages 231–239. Springer Berlin Heidelberg, 2013.

[234] Paul Syverson, Catherine Meadows, and Iliano Cervesato. Dolev-Yao is no better than Machiavelli. In P. Degano, editor, *Proceedings of the First Workshop on Issues in the Theory of Security - WITS'00*, pages 87–92, Geneva, Switzerland, 07 2000.

[235] Herman T. Tavani and James H. Moor. Privacy Protection, Control of Information, and Privacy Enhancing Technologies. *ACM SIGCAS: Computers and Society*, 31(1):6–11, 2001.

[236] M. Tehranipoor and F. Koushanfar. A survey of hardware trojan taxonomy and detection. *Design & Test of Computers*, 27(1):10–25, 2010.

[237] Ken Thompson. Reflections on Trusting Trust. *Communications of the ACM*, 27(8), 1984. *ACM Turing Award 1984 acceptance speech paper.*

[238] Iain Thomson. Instagram back-pedals in face of user outrage, 2012.

[239] Rupert Thorogood and Charles Brookson. Lawful Interception. *Telektronikk*, 103(2):33–36, 2007.

[240] TrueCrypt Developers Association. Free open-source disk encryption software, 2013.

[241] J. Vaidya and C. Clifton. Privacy-preserving data mining: why, how, and when. *Security Privacy, IEEE*, 2(6):19–27, 2004.

[242] Jaideep Vaidya and Chris Clifton. Privacy preserving association rule mining in vertically partitioned data. In *Proceedings of the eighth ACM SIGKDD international conference on Knowledge discovery and data mining*, KDD '02, pages 639–644, New York, NY, USA, 2002. ACM.

[243] Jaideep Vaidya, Murat Kantarcioglu, and Chris Clifton. Privacy-preserving naive bayes classification. *The VLDB Journal*, 17(4):879–898, July 2008.

[244] Marten Van Dijk and Ari Juels. On the impossibility of cryptography alone for privacy-preserving cloud computing. In *Proceedings of the 5th USENIX conference on Hot topics in security*, HotSec'10, pages 1–8, Berkeley, CA, USA, 2010. USENIX Association.

[245] Serge Vaudenay. *On Privacy Models for RFID*, volume 4833 of *LNCS*, pages 68–87. Springer, 2007.

[246] Charles A. Walton. Electronic identification & recognition system. US Patent 3,752,960, 08 1973.

[247] Charles A. Walton. Portable radio frequency emitting identifier. US Patent 4,384,288, 05 1983.

[248] Samuel D. Warren and Lois D. Brandeis. The Right to Privacy. *Harward Law Review*, IV(5), 12 1890.

[249] Gary M. Weiss. Data mining in telecommunications. In Oded Maimon and Lior Rokach, editors, *Data Mining and Knowledge Discovery Handbook*, pages 1189–1201. Springer US, 2005.

[250] Gilbert Wondracek, Thorsten Holz, Engin Kirda, and Christopher Kruegel. A practical attack to de-anonymize social network users. In *Security and Privacy (SP), 2010 IEEE Symposium on*, pages 223–238. IEEE, 2010.

[251] Yangyang Wu, Shuguang Du, and Wei Luo. Mining alarm database of telecommunication network for alarm association rules. In *Dependable Computing, 2005. Proceedings. 11th Pacific Rim International Symposium on*, page 6 pp., 2005.

[252] Li Xiong, Subramanyam Chitti, and Ling Liu. k nearest neighbor classification across multiple private databases. In *Proceedings of the 15th ACM international conference on Information and knowledge management*, CIKM '06, pages 840–841, New York, NY, USA, 2006. ACM.

[253] A.C. Yao. Protocols for secure computations. In *Proceedings of the 23rd Annual Symposium on Foundations of Computer Science*, pages 160–164, 1982.

[254] V. Zadorozhny, V.A. Oleshchuk, and P. Krishnamurthy. A framework for efficient security and privacy solutions in data intensive wireless sensor networks. *TELEKTRONIKK*, 103(2):61, 2007.

[255] Rui Zhang, Yanchao Zhang, and Kui Ren. Distributed privacy-preserving access control in sensor networks. *IEEE Transactions on Parallel and Distributed Systems*, 23(8):1427–1438, 2012.

[256] Gansen Zhao, Chunming Rong, M.G. Jaatun, and F.-E. Sandnes. Deployment models: Towards eliminating security concerns from cloud computing. In *High Performance Computing and Simulation (HPCS), 2010 International Conference on*, pages 189–195, 2010.

Index

About the Authors

Geir M. Køien is an Associate Professor of Computer Science at University of Agder, Norway. He received his PhD. degree in Communications (2008) from Aalborg University, Denmark. He is a senior member of the IEEE and a senior member of the ACM. His current research interests include system architecture and access security, privacy and trust with special focus on telecommunication systems. He has previously worked for Ericsson Norway and Telenor R&D with research and development within the 3GPP mobile systems, and he was the Telenor delegate to the standardization work in the 3GPP SA3 (Security) work group for 10 years. He received the Best Paper Award in ACM WiSe in 2005 and in MobiSec in 2010.

Vladimir A. Oleshchuk is a Professor of Computer Science at University of Agder, Norway. He received his Ph.D. degree in Computer Science (1988) from the Taras Shevchenko Kiev State University, Kiev, Ukraine. From 1987 to 1991, he was Assistant Professor and then Associate Professor at the Taras Shevchenko University. He is employed at the University of Agder since 1992, first as Associate Professor, and then as Full Professor. He is a senior member of the IEEE and a senior member of the ACM. He has served as a program committee chair, program committee member and reviewer for many international conferences. His current research interests include formal methods and information security, privacy and trust with special focus on telecommunication systems and eHealth. He received the Best Paper Award in IEEE MDM 2012.